Biometric System and Data Analysis
Design, Evaluation, and Data Mining

Biometric System and Data Analysis

Design, Evaluation, and Data Mining

by

Ted Dunstone
Neil Yager

Eveleigh, NSW, Australia

 Springer

Editors:

Ted Dunstone
Biometix
National Innovation Centre
The Australian Technology Park
Eveleigh, NSW 1430
Australia
ted.dunstone@biometix.com

Neil Yager
Biometix
National Innovation Centre
The Australian Technology Park
Eveleigh, NSW 1430
Australia
neil.yager@biometix.com

ISBN-13: 978-1-4419-4595-2   e-ISBN-13: 978-0-387-77627-9

Printed on acid-free paper

springer.com

Preface

Overview

Biometrics is the identification of an individual using a distinctive aspect of their biology or behavior. Biometric systems are now being used for large national and corporate security projects, and their effectiveness rests on an understanding of biometric systems and data analysis.

Books on biometrics tend to focus on the details of biometric systems and their components, and distinguish between the various biometric modalities. This book is different. It presents a unified view by focusing on the common aspects of all biometric systems - the input (biometric images and person metadata) and the output (similarity scores). The matching algorithms and sensing technologies may change with each new advance, however the decision process for matches is not affected. Building on this base, this book brings together core aspects of statistics and probability to provide a comprehensive guide to evaluating, interpreting and understanding biometric data.

In doing so, we paint a coherent and intuitive picture which will be equally useful to novices and advanced readers. All theoretical concepts are grounded with practical examples and techniques for the evaluation and set-up of real-world, operational biometric systems. Along the way, other relevant topics are introduced - including machine learning, data mining and prediction - which have been widely applied to other fields, but not yet rigorously applied to biometrics. Case studies and examples from several major biometric modalities are covered.

The focus of biometric research over the past four decades typically has been on the bottom line of driving down system-wide error rates, and as a result powerful recognition algorithms have been developed for a number of biometric modalities. These algorithms operate extremely well under laboratory conditions, but their performance can fall short in the real-world. One reason has been the focus on algorithmic performance as measured by collective statistics. A major theme of this book is the importance of data quality and the accuracy variation between individual users.

These factors form the basis of system performance, and understanding their effects will be necessary for the development of the next generation of biometric systems.

Before putting any biometric system into operation, it needs to be evaluated for accuracy, security and effectiveness. A variety of performance measures are available, and selecting an inappropriate measure can result in highly misleading statistics. This book places an emphasis on the various biometric performance measures, what they mean, and when they should and should not be applied. The evaluation techniques are presented rigorously. However, they are accompanied by intuitive explanations that can be used to convey the essence of the statistical concepts to a general audience.

Objectives

This book aims to provide a comprehensive treatment of understanding and improving biometric systems at various levels. Our practical experience in evaluating and working with a wide range of real-world biometric systems and data is used to inform newcomers and experienced practitioners through clear diagrams and carefully prepared examples. A number of novel techniques for biometric analysis are also presented. The book provides a solid understanding of:

- The fundamentals of biometric systems
- Organization and structure of biometric information
- The basics of multimodal systems
- Biometric data quality issues and their impact on system performance, in particular pictorial examples for fingerprint, face, iris and speech
- The assessment of individual user and user group accuracy
- Setting up and conducting biometric evaluations (there are dedicated chapters on identity document and surveillance systems)
- ISO-consistent vocabulary for all descriptions
- Theoretical methods to create standard system level statistical measures
- Techniques to establish the existing level of fraud detection in identity systems
- Assessing biometric vulnerabilities

Audience

This book represents an effort to bridge the divide between biometric researchers and the biometric industry. It aims to appeal to people with an operational responsibility for biometric systems, as well as technical and academic readers. In general, Part I is directed towards people with operational responsibility and Part II towards technical experts and academics. However, it is hoped that the book as a whole will be of benefit to both audiences. Introductory readers can use the more detailed aspects in Part II as a reference and a source of information once they have a firm understanding of the basics. Readers with a strong technical background can quickly gain a high-level, intuitive view of the field from Part I. All readers will gain valuable insight into the nature of biometric systems from the examples throughout the book, which are the result of many years of practical experience.

The more technically challenging chapters in Part II - the biometric performance hierarchy - have simple and clear introductory sections. Sections marked with a '*' include more complex, mathematically-oriented material which can be skipped without losing continuity. Diagrams and examples have been extensively used to illustrate the techniques. The emphasis is on practical analysis techniques that are useful for analyzing a variety of real biometric systems over different modalities.

Appealing to introductory and advanced technical readers in the same volume risks making the scope so wide that neither group's needs are adequately addressed. We hope to have avoided this pitfall by drawing from a variety of sources, both academic and industrial.

The following are specific audiences that have been identified, along with the chapters they are likely to find particularly relevant:

- *System Implementers and Designers*: A detailed understanding of techniques helps make sure a system has been configured and tuned correctly. The detection and diagnosis of common problems is a major theme of this book. In particular Chaps. 1, 2, 3, 6 and 12 contain useful information.

- *Researchers and Students*: Every effort has been made to ensure students are guided through both the introductory and advanced topics. For new students, working through Part I should give a good basic grounding in biometrics and biometric analysis. Researchers will be particularly interested in the techniques presented in Chaps. 8 and 9, which present a novel, user-centric framework for system analysis.

- *Security Consultants and System Evaluators*: Accuracy evaluation of system performance and the explanation of results to stakeholders are difficult and costly. This book aims to ensure that the outcomes from an evaluation actually measure expected performance and can be communicated clearly to a non-technical audience. Chaps 1, 2, 3, 5 and 6 are particularly relevant.

- *Forensic Investigators*: For anyone using a biometric system for forensic investigation or legal purposes, it is vital to have a solid understanding of the likelihood of a match being the actual person concerned in order to provide a statistical justification for a decision. In addition to the introductory chapters of Part I, the specialized chapters on identity systems (Chap. 10) and surveillance (Chap. 11) will be valuable.

- *Surveillance Operators*: Setting up and evaluating biometric surveillance systems is more difficult than for other biometric systems. Chapter 11 presents techniques for designing these systems and running trials for their evaluation.

- *Vendors and Algorithm Developers*: This book should allow those developing biometric systems to undertake more realistic testing and evaluation, and hence gain a better understanding of factors that impact on real-world performance.

The measures introduced in Parts II and Part III should be of primary interest in providing new tools and insights into how to improve the accuracy of existing algorithms.

- *Auditors:* As biometrics become part of larger and more critical systems, the need to independently audit their performance is increasing. Traditional IT security auditing techniques do not adequately encompass the non-deterministic nature of the security outcomes for biometric systems. The techniques outlined in Part I will give any auditor a good idea of the issues that need to be considered for auditing a biometric system. In addition, Chap. 12 will be of assistance in assessing a risk management plan.

Organization

This book is in three parts: a general introduction to biometric fundamentals, a detailed technical treatment of analysis techniques and special topics in data analysis.

Many core concepts are presented twice. They are introduced in Part I, with an emphasis on examples and intuitive understanding. Part II provides a deeper, more technical and statistically grounded analysis. Each chapter can be read on its own, but there is a clear progression of ideas from introductory to advanced and from generalized to specific. The terminology in Chap. 6 has been written to be consistent with the ISO 19795.1 standard.

The content is:

Part I: Preliminaries provides a step-by-step introduction to biometric systems, data and analysis that does not require any statistical background. Chapter 1 introduces biometric systems, how they work and related issues such as privacy and vulnerability. Chapter 2 provides the preliminaries of the analysis of biometric systems and data through simple diagrams and worked examples. This chapter has been written to ensure that even readers who are not familiar with statistical techniques can follow the descriptions and gain an appreciation of how biometric systems make decisions. Chapter 3 looks at the biometric data. Practical examples are given for fingerprint identification, face recognition, speaker verification, iris recognition, vascular recognition, keystroke dynamics and others. The implicit structure and organization of biometric matching results are illustrated, along with references on how they can be efficiently stored and retrieved. Multimodal systems, the use of more than one biometric matching algorithm or characteristic, are examined in Chap. 4. In Chap. 5, general requirements for running a biometric evaluation are introduced along with practical considerations and guidelines. Chapter 6 introduces the standard terminology used throughout the book, which is consistent with standard ISO definitions and is a convenient reference.

Part II: The biometric performance hierarchy examines in detail the analysis of biometric systems, from the highest level of system evaluation to the performance of individual users. Chapter 7 details the statistical basis of biometric systems. Starting from the fundamental building block of a single match score, collective error rates for the system as a whole are developed. Verification, closed-set identification and open-set identification systems are differentiated, with an emphasis on the specific measures that are appropriate for each type of system. Almost always, some users of biometrics systems perform better than others. These differences are the subject of Chap. 8. There is a description of a framework which allows the detection and characterization of problem users, based on individual evaluation and the members of the biometric menagerie. Chapter 9 presents the novel application of data mining techniques to biometric systems to extract knowledge from masses of information. In particular, machine learning is used to automatically detect groups of problem users with common attributes.

Part III: Special topics in biometric data analysis introduces specific topics on the analysis of biometric data and systems. Chapter 10 is on the topical subject of identity document identification systems, identity theft and proof of identity. Techniques to estimate the level of fraud in existing biometric databases are presented, as well as a discussion on the use of biometrics in a legal setting. The set-up and analysis of covert surveillance systems is examined in Chap. 11. The analysis of data from surveillance systems is complicated by the less structured and less controlled environments where they are often required to operate. Chapter 12 provides an introduction to detecting and mitigating vulnerabilities in biometrics systems.

Subjects Not Covered in Detail

This book is different from other books on biometrics. By placing an emphasis on biometric data and analysis, some subjects have been covered in less detail:

- The emphasis is on the input and the output of biometric algorithms; not what happens in between. In other words, the actual matching engine is treated as a "black box". The book does not cover in detail the underlying image-processing and pattern-recognition algorithms which form the core of biometric matching. Also, the hardware and sensor mechanisms used for the acquisition of biometric samples are not detailed, except where it is relevant to the analysis. The analysis is picked up once a similarity score has been generated and it is followed all the way to high-level performance measures.
- There is no attempt to comprehensively review emerging biometrics. However, the techniques introduced are applicable to any new biometric type and, indeed, could be applied to many other areas of pattern classification, from medical imaging to number-plate recognition.
- In general, the focus is on quantitative rather than qualitative analysis. Qualitative aspects of a biometric system include considerations such as cost and ease of use, and these are not discussed in detail.
- In some areas, such as confidence intervals, there are still open questions and academic debate about the best techniques. When this is the case, we present only an overview of the most common techniques. In other areas, such as multimodal

analysis, techniques still are emerging and are treated only at an introductory level in this edition.

Original Research

This book presents a number of topics which have not been comprehensively considered in previously publications. Some insights are the results of original research we have conducted into the analysis of biometric systems, while others have grown naturally from our practical experience in the field.

Original research covers:

Zoo Analysis   Through our efforts in evaluating real-world systems, we have gained an appreciation of the importance of the performance of individual users of biometric systems. The original members of the biometric menagerie (sheep, goats, lambs and wolves) are well known. However, we have recently extended this group of animals to characterize other problem users. Chapter 8 contains a detailed analysis of the subject, as well as presenting a user-centric framework for system evaluation. The zoo analogy is used extensively and is an integral component of any biometric evaluation.

Data Mining   Although user-level analysis has achieved a level of acceptance in the biometric community, the existing techniques for group-level analysis are primitive. In general, trends in the data are currently discovered through a manual process of directly computing and comparing group accuracy (i.e. men vs women). Chapter 9 draws from the fields of machine learning and data mining, and presents the novel application of intelligent techniques for automated biometric knowledge discovery.

Fraud Level Identification   One of the most promising applications of biometric matching is detecting fraud in identity databases. For example, driver's license and passport authorities often have little concrete evidence for their estimates of existing levels of fraud. Chapter 10 presents the results of our initial investigations in this area. In particular, it outlines tests that need to be run, and how the results can be used to establish the extent of fraudulent activity.

Surveillance Systems   We have particular expertise in the analysis of biometric surveillance systems. This is an emerging application and there is little published information on the best ways to design and assess these systems. In Chap. 11 we highlight the major issues and provide practical tips for setting up systems and running evaluation trials. This chapter will be an invaluable resource and time-saver for anyone involved in surveillance evaluations.

Vulnerabilities   Accuracy in biometric systems has traditionally been based on the probability of success for a random, non-motivated impostor attempting to gain system access. The discovery and reporting of vulnerabilities in biometric systems has not been particularly well analyzed, even though it has a crucial role in ensuring a secure system.

Acknowledgments

This book is the culmination of many years working with, and evaluating, biometric systems. Along the way, correspondence with leading practitioners such as Jim Wayman and Tony Mansfield has contributed significantly to our knowledge, and for this we are greatly indebted. Other colleagues who have shared valuable ideas with us over the years include Johnathon Phillips of NIST; Michael Petrov, Michael Brauckmann and Jonathan Wells of L1; Alfredo Herrera, Frank Weber and Raphael Villedieu of Cognitec; Geoff Poutlon, formerly of CSIRO; Aidan Roy of the Institute for Quantum Information Science; Terry Hartmann of UNISYS; Dijana Petrovska of Biosecure; Terry Aulich of Aulich & Co; Valorie Valencia and Roger Cottam of Authenti-Corp; Brett Minifie and Ian Christofis of HP; and Stephen J. Elliott and Eric Kukula of Purdue University.

Over the years, the clients of our consulting company, Biometix, have challenged us to be clear in our thinking and explanations - our discourse with them has increased our own understanding, ultimately allowing us to write this book with clarity. In particular, it has been a pleasure to work with Ross Summerfield of Centrelink; Kenneth Beaton of KAZ; Ossie Camenzuli, formerly with the RTA; Dominique Estival, Stephen Anthony, Stephen Norris and Son Bao Pham of Appen; Jason Prince, Johnathan Moulds and Karen Shirely of the AFP; and Rick Hyslop and Derek Northrope of UNISYS. Thanks also to Barry Westlake for his inspiring vision of the future of Biometix, and to Fabrice Lestideau, Teewoon Tan and Navin Keswani for their help in building the foundations. A great many others from the research, commercial and government communities, particularly those who are part of the Biometrics Institute, have shaped our appreciation for the many different aspects of biometrics, and to all those we express our deep gratitude.

Min Sub Kim has provided invaluable feedback, and a thorough proof reading, for Part II of this book. Furthermore, editing done by Patrick Weaver, and the huge efforts by Coralie and Simon Dunstone, have contributed greatly to ensuring a high standard and consistent quality throughout. Despite the monumental efforts of all those who helped in the preparation of this manuscript, the authors take full responsibility for any mistakes, inconsistencies or omissions.

Data sets, examples and diagrams were kindly provided by Johnathon Phillips, Tony Mansfield, David Cho, Sammy Phang and Patrick Flynn.

Ted would like to thank the support and forbearance of his friends, the people from the Institute, Isabelle Moeller and Michelle Turner, and in particular the patience (and fabulous food) of Susan Crockett while this book was written - often at odd hours. Neil would like to thank his family for being by his side, even from a world away. He also thanks Fjóla Dögg Helgadóttir, who sparkles and shines, for helping in ways that words cannot measure. This book, and much besides, would not have been possible without her smile.

Ted Dunstone
July 2008   *Neil Yager*

There are a number of sections of this book where the reader may find it beneficial to perform some worked examples using computer software.

The authors have made data and a limited license to a computer program available, _free of charge_, for readers and their associates, that will make it easy to perform these examples.

This data and the limited license to PerformixPC can be downloaded from www.biomet.org using the special code "PRMDEMO".

This code is not unique so it may be given to others, including students, to carry out these examples.

The limited license for PerformixPC provides virtually all of the functions of the software and does not expire, but is only able to operate on small data sets such as those used in these examples.

This software and data is supplied without any guarantees of its fitness for use and the authors and the publisher accept no liability for its use.

Contents

Part I
An Overview of Biometrics

Biometric systems often fail not because of the underlying matching technology, but as the result of an inadequate set-up and configuration leading to poor operational performance. Partly, this is a consequence of set-up processes that have tended towards a black-art involving significant trial and error, rather than a systematic search for the best parameters.

A significant portion of this book has been inspired by a desire to ensure practitioners know when their biometric system has been set-up and configured correctly. Much of the existing literature in this area is directed more towards researchers than to implementers, integrators or managers. Consequently, there often is a gap in understanding between the two groups about the underlying principles of biometric systems.

The primary aim of Part I is to explain, in a way that does not require a deep statistical understanding of the subject, how biometric systems work and how they are configured. However, this is not done at the expense of rigor or correctness. The results of a biometrics evaluation are easy to misrepresent or accidentally misinterpret, even for experienced practitioners. Therefore, it is important that common-sense explanations and clear illustrations are used to detail the interpretation of results, as outlined in Part I of this book.

Chapter 1
An Introduction to Biometric Data Analysis

Biometrics is a fascinating field. What other area of science or engineering combines aspects of biology, statistics, forensics, human behavior, design, privacy and security - and also spans everything from the simple door lock to huge government systems? This diversity is compounded by the wide biological and behavioral variation in people, making it an intriguing challenge to evaluate, configure and operate a biometric installation. However, regardless of the biometric type, the fundamentals of how match decisions are made are common to all biometrics. These unifying features allow an introduction and discussion of biometrics through a common framework.

This chapter provides a snapshot of biometric systems and analysis techniques. It previews the contents of the rest of the book, providing a concise introduction to all the main topics. The goals of this chapter are to:

- Introduce the history of biometric science (Sect. 1.2).
- Discuss how biometrics fit into an identity management framework (Sect. 1.3).
- Discuss desirable aspects to consider when choosing a biometric (Sect. 1.4).
- List the components and forms of biometric data (Sect. 1.5).
- Provide an overview of biometric systems (Sect. 1.6).
- Give high level descriptions of the major biometric graphs (Sect. 1.7).
- Introduce aspects of biometric privacy (Sect. 1.8).

1.1 Introduction

Biometrics is the use of distinctive biological or behavioral characteristics to identify people. Automated biometric recognition mimics, through a combination of hardware and pattern recognition algorithms, a fundamental human attribute - to distinguish and recognize other people as individual and unique. There is a long history of using distinguishing marks for identification. From hand-prints on cave walls and hand-written signatures on manuscripts, to the direct measurement of head dimensions and the unique patterns of Morse operators, there has long been a desire to use and measure biometric identity.

This book is concerned with the automation of this process, where a computerized system can make a determination on identity without human supervision. This field is called biometrics and its path from research to wide-ranging adoption is still unfolding. Huge advances have been made in every aspect of biometrics since the first automated matching systems in the late-1970s, from the underlying technology through to the increasing number of biometrics in everyday use.

Systems incorporating biometrics now span the globe in a variety of applications from law enforcement and passport control to laptops. The potential impact of biometrics is enormous, as demonstrated by the numerous areas of authentication using traditional security, such as keys, passwords and cards. Biometrics will be an increasingly integral part of the solution to challenges from rising global levels of identity theft to the demand for convenience and security. Areas as diverse as health care, travel, call centers, banking, Internet security, cars, consumer electronics, voting, proof of identity and payment systems are all starting to see the benefits of the application of biometrics.

One principal advantage is the increased confidence it brings to transactions. This happens because when a biometric is used, the person must have been physically present during the authentication. Unlike a password or keys, biometrics cannot be given to another person. This is often a principal driver for its adoption in the government and commercial spheres. The other significant benefit is convenience - often it is simply faster and easier to use a biometric sensor. It is this factor that usually drives the growth in the consumer area. Both security and convenience can lead to reduced costs, since riskier transactions can be undertaken with less supervision, and transactions are less like to fail because of forgetfulness.

This book is the result of over 15 years of analysis of government and commercial biometric systems, encompassing many different real-world evaluations from a wide variety of activities. Through this experience, techniques and insights have been developed to analyze biometric systems, which should contribute to making biometrics operate better and more accurately.

It has been a necessary part of our work to ensure that those who need to implement or manage biometric systems understand how they make verification decisions and how to interpret performance graphs. No matter if you are a newcomer or an experienced practitioner, we hope this book will broaden your understanding of biometric systems and inspire you to explore this exciting area further. In this chapter, the construction and operation of biometric systems is introduced.

1.2 A Brief History of Automated Biometric Identification

The manual identification of people before the introduction of computers is extensively discussed in a number of other books [4, 7, 16], but the subsequent development of biometrics is the fascinating story of the moment. Serious biometric research began in the 1960s, the techniques were developed and refined during the 1970s and 1980s, and the field became increasingly commercialized from the mid-1990s onwards.

1.2.1 The Hands: Fingers, Palms and Hands

The initial demand for automated biometric analysis was driven by the need to process *fingerprints* in the criminal justice system. This task historically required an experienced practitioner to examine fingerprint images and classify them into different types based on the overall pattern of ridges and valleys. These were then manually compared to establish whether the suspect had been seen previously or if there was a match to a crime scene. When the number of fingerprints held by the police grew massively in the early 20th century, the manual process of classifying and searching for fingerprints became prohibitively costly and error prone.

It was not until the early 1960's, and the advent of powerful computers, that the first experiments in the biometric matching of fingerprints were conducted at the U.S. National Bureau of Standards [7]. In 1979 the first working prototype of a fingerprint searching system was tested at the U.S. Federal Bureau of Investigation. By 1983 Automated Fingerprint Identification Systems (AFIS) were in routine use, and within three years they were being adopted globally. This set the scene for the large scale growth of the biometrics industry.

Palm-prints are now also used in AFIS systems. The first reported commercial palm-print system came from Hungary in 1994, and by 1997 similar processes were being built into other AFIS systems.

In the mid-90s, experts developed fingerprint sensors that were cheap enough and small enough to be used for access control and computer login. At first these were optical systems using similar technology to AFIS. Around this time it was accidentally discovered that fingerprints could be read straight from their capacitive effects on silicon wafers. This led to a drastic reduction in size and cost of sensors, and consequently the market for sensors for laptops and other consumer products experienced enormous growth.

Advances in Fingerprint Scanners

In the mid-90s a Bell Labs employee was, according to legend, experimenting with a DRAM that was getting hot. When he put his finger on to check the heat he discovered that he had flipped the bits in the memory registers where his fingerprint ridges had been. The first company to commercialize this discovery was Veridicom. The next significant innovation was when the industry moved from large "placement" sensors at $50 each to swipe sensors of around at $5 each. Future developments should continue to reduce swipe sensor costs to $2 to $3 whilst not compromising reliability. Swipe sensors also enable dual functionality as they can be used as part of the user interface, as well as providing security on mobile phones and laptops.[a]

[a] Thanks to Brett Minifie from Hewlett-Packard Australia for this information.

Technology	Approximate date of an early major paper or relevant patent	Approximate date of an early commercial implementation
Fingerprint (AFIS)	1962 paper	1979, 1985 Identix
Retina	1978	1984, EyeDentify, Inc.
Speaker	1963 paper, Pruzansky	1976, Texas Instruments
Face	1965, Helen Chan and Charles Bisson	1996 Cognitec,ZN, Identics
3D Face	1992 G. Gordon	2001, A4 Vision
Hand	mid-1960s	1986, IR Recognition Systems
Iris	1987 Patent, John Daugman	1995, Iridian
Palm	1994	1997
Vascular	1992, Dr K. Shimizu	early-2000s
Finger Vein	2002	2004
Keystroke	1986 Patent J. Garcia.	2002 iMagic

Table 1.1 History of some biometric developments. The events listed relate to the first significant developments in research and commercial adoption of automated identification.

An early example of a successful commercial biometric used for access control was *hand recognition*, based on the geometry of the hand. The original technology was developed in the mid-1960s, and started to enjoy wide commercial success in 1986. At the 1996 Atlanta Olympic Games it was used as part of the security access control for athletes. Compared with its main competitors in access control – fingerprint and iris recognition – this technology requires large, bulky readers and, as a consequence, now has a declining market share.

The individual pattern of veins in the fingers, wrist, hands or face can be mapped using infrared cameras. This is known as vascular recognition. Because the sensor is non-contact, the readers can be made quite robust, and it is more difficult to covertly obtain these patterns than many other biometrics. Since 2005, the popularity of finger and palm vein scanners has been growing, particularly in Japan where they are being used in automated teller machines.

Art, Literature and Biometrics

Prehistoric artists used hand-prints in cave paintings, perhaps as a 'signature'. They might be considered the earliest example of a biometric identifier. Such hand-prints are found in at Lascaux in south western France and date from at least the Upper Paleolithic period (approximately 16,000 years ago).

Today's artists also are looking to mark their work. Recently, the Australian artist ProHart used DNA from a cheek swab and mixed it with paint to uniquely identify his paintings. Similar DNA labeling techniques, coining the term "spit label", have been used to create an indigenous 'authenticity label' to protect artists from fraudulent copying [3].

Biometric techniques can also be used to uncover forgeries. Famous examples include the fake painting "Skating in Holland" 1890-1900, which is signed by Johan Barthold Jongkind. However, the signature on the painting is not the same as the artist's real signature. As well, signal processing techniques can be used to pick up the inherent style of an artist that is embedded in the texture of the images. When the painting "Virgin and Child with Saints" by Pietro Perugino was analyzed, the work appeared to be from at least four different artists.

In literary works, Shakespeare's authorship of some plays has been questioned since as far back as the 18th century. Stylistic analysis of plays (called stylometry) known to have been written by Shakespeare provides a behavioral biometric signature of the way he used words. The analysis suggests that some works were collaborations (in particular, the plays "The Two Noble Kinsmen" and "Henry VIII" which were co-written with John Fletcher).

1.2.2 The Head: Face, Voice and Eyes

Because people commonly use faces for establishing identity, it is perhaps the most natural biometric for authentication. However, it is also one of the most challenging for a computer. The earliest work published in the automation of *face recognition* was in 1965 by Helen Chan and Charles Bisson. A system was developed by a small research company (called Panoramic) that looked at manually marked points on a face, and used the marked points to search a database of several thousand people [5]. Automated face recognition system research was undertaken by Kanade in 1977 [12], but it did not become a significant research area until the 1990s. The seminal papers that gave rise to fully automated face recognition came first in the late-80s and early-90s [13, 14, 18], and these showed that faces could be represented using only a few hundred parameters.

The first modern commercial face recognition systems began to appear around the mid-90s. Amongst these early face recognition companies were Cognitec, ZN, Viisage Technology and Visionics Corporation. By 2006, L1-Systems had been formed through the merging of ZN, Viisage Technology and Visionics Corporation.

Advances in technology continue to provide significant performance improvements. One of the most significant was the use of the skin texture analysis, introduced in 2004. This used the high-frequency information encoded in the blemishes and skin pores to boost recognition accuracy for high-resolution images.

In 1963 a paper was published on "Pattern matching procedure for automatic talker recognition" [21]. The first early *speaker verification system* subsequently came out of AT&T's research labs in 1974 [11] and in 1976 Texas Instruments built a prototype system that was tested by the U.S. military. Through the 1980s and 1990s, steady research progress was made. This meant systems could operate under text-independent and text-dependent modes and increased the robustness of the processing algorithms. One of the seminal papers for the analysis of biometric data was published in 1998 on the classification of user types in speaker verification, known as Doddington's Zoo [9]. Examples of the user types identified included goats (people who had trouble being recognized) and wolves (people who were naturally able to sound like others). This is the original basis of research and analysis techniques presented on individual evaluation in Chap. 8.

A biometric which has started to be adopted widely over the past 10 years is the unique pattern found on the iris - the colored muscle in the center of eye. Two ophthalmologists, Dr. Leonard Flom and Dr. Aran Safir, patented the idea in 1987 and worked with Prof. John Daugman to develop an *iris recognition* algorithm. By 1994 Daugman had received patents for his algorithms. Successful tests for the U.S. Defense Nuclear Agency were completed the following year along with a commercial implementation. Iris recognition was proposed in 2003 as a part of a potential British national identity card.

In contrast to the iris, which is the front of the eye, *retina recognition* uses of the pattern of the blood vessels at the back of the eye. Its first commercial implementations were in the mid-1970s. The equipment for this was large and expensive and there were difficulties identifying people wearing glasses.

1.2.3 Other Biometrics

Most people's first exposure to the concept of a biometric identifier is through a handwritten signature. In many cases, a signature by itself is not very distinctive, can be highly variable and is relatively easy to copy. Because of this, research has concentrated on the use of dynamic features for online recognition, such as pressure and acceleration.

Keystroke dynamics were first recognized as a biometric identifier during World War II. Allied telegraph operators could identify other operators, friendly and enemy, through a keying rhythm which was called the "The Fist of the Sender". Even without decoding messages, this allowed German troop movements to be monitored, because the German telegraph operators were usually attached to a particular fighting unit. Modern keystroke dynamics uses the timing difference between keystrokes and looks for idiosyncrasies in the use of keys (e.g. how long the typist holds down the shift key or uses the control keys). Commercial keystroke dynamic systems were available in the early-2000s, but they have not been widely adopted because they lack accuracy and require long training times. However, more sophisticated systems have been recently proposed that claim better performance.

Even the way people walk, their gait, has been used as an identifier. Research on this has been prompted by the need to identify people from surveillance cameras at a distance in poor light, without a face clearly visible. This is a difficult task since a person's gait is highly variable and clothing obscures fine differences.

The field of biometrics continues to develop new methodologies for recognition. For instance, in 2008 a patent was even granted for "Method and system for smell-print recognition biometrics on a smart-card".

1.2.4 Post September 11, 2001

The terrorist attacks in the U.S. on September 11, 2001 focused enormous attention on biometrics, and on its ability help secure the U.S. and other countries against terrorist threats by making it more difficult to use fake identities. The 9-11 Commission report, produced in the wake of the attacks, suggested the introduction of biometrics at national borders to prevent unwanted foreign nationals, particularly people traveling under a false identity, from entering the country. The UN body controlling passport standards, the International Civilian Aeronautical Organization (ICAO), for some time had been looking at a new standard for electronically-readable passports that included biometrics. After 2001, this work took on renewed urgency and by early 2008 many countries had implemented passports that include biometrics.

The mandatory biometric in the passport standard (known as IACO/MRTD Doc 9303) was a facial biometric token, as this was most compatible with existing passport-issuing infrastructure and was the most widely acceptable. Fingerprint and iris recognition are optional biometrics that could be included on the passport and have been used by a variety of countries.

The result of the terrorist attacks on the biometrics industry was mainly long term. In the short term, biometric trials were run and discussion papers produced, but many large scale implementations took years to appear. Significantly, however, more money was diverted into research, and this focus is now resulting in substantially improved accuracy and usability.

Legitimate Fake Identities

An interesting side effect of the adoption of biometrics has been the difficulty for covert operatives, or those in witness protection, to obtain false identities without being detected by the issuing agencies. In essence, the person's biometric ties them to their true identity, and an alert may be raised to the operator when they try to obtain an ID under a different name. Under previous circumstances, these agencies would have no idea a person had multiple identities, legitimate or otherwise.

Biometrics also had problems with public perception in the post 9-11 world. In particular, it is commonly being seen as a big brother technology, potentially leading to the invasion of privacy. Civil liberty and consumer groups have been concerned about the implications of large biometric databases and the potential for misuse of this data through its sale to third parties or its use for unauthorized purposes (such as cross matching with other biometric databases). These concerns were first raised in a significant public way when face recognition was used to scan the crowds at the 2001 football Super Bowl in Tampa, Florida.

Substantial improvements to sensors, algorithms and techniques are likely over the next 10 years. In particular, we can expect continuing advances in the biological study of how brains make complex decisions, and the development of new nanotechnology materials which will transform biometric technology. Regardless of these advances, the fundamentals of biometric data analysis discussed in this book are unlikely to change.

1.3 Identity and Risk Management

Biometrics is often used to control the risk of a security breach and to facilitate convenient transactions. In this context, biometrics is part of the risk and identity management process. Identity management encompasses all phases of the process of dealing with information relating to an individual's identity. This involves regulating and controlling the places where identity information is used or processed, extending from the initial identity creation and verification, through their use and potential reissue, to the final removal of the identity.

It is particularly important in any large biometric system that the full identity life-cycle is supported. A biometric alone will never add security to a system that does not have good identity controls around the initial, or subsequent, enrollment of individuals.

Where anyone can self-enroll a biometric with no check or audit, system circumvention is easy, regardless of the strength of the biometric control. This is because there is no way to ensure the correct identity is bound to the enrolled biometric. The binding of an identity to a biometric is achieved by authenticating the supporting documentation, such as birth certificates or passports. This is commonly known as proof of identity.

Extreme Biometric Makeovers

Several cases have been reported of people changing their biometrics to create new identities - usually with disastrous consequences. In 1997, Amado Carrillo Fuentes, one of the major drug traffickers in Mexico, underwent facial plastic surgery and liposuction to change his appearance. He died as a result of complications. This is perhaps the first, and hopefully last, death of someone specifically trying to create new biometrics. In 2007, a Mexican doctor was arrested in the U.S. on suspicion of trying to change a drug dealer's fingerprints by surgically replacing them with skin from the bottom of the feet. It was reported that the operation left the drug dealer's fingers barely usable.[a]

[a] http://www.iht.com/articles/ap/2007/05/11/america/NA-GEN-US-Mexico-Fingerprint-Removal-Charges.php

---

Understanding biometric statistics and their impact on risk management can be complex, particularly for large public-facing systems. This is because the performance of biometrics is highly dependent on the characteristics of customers and the environment where the system is being used. For instance, with some biometric systems, a percentage of users always will have difficulty with the biometric sensor for biological, cultural or behavioral reasons. Techniques can be implemented to manage this but care must be taken to ensure customers do not feel disenfranchised, and that the overall level of security is not diminished. Consequently, it is important to undertake a proper risk assessment, incorporating a detailed understanding of the risk, threats and opportunities provided by biometrics, and follow up with regular biometric audits.

1.4 Desirable Biometric Attributes

There is a huge diversity of biometrics types, and each biometric has its own particular strengths, depending on the application. Underlying this diversity are some general attributes that are desirable for good biometric operation [16]. The importance of each of these factors depends on the results required for a particular project.

Where biometric systems are being compared as part of a formal selection process, each attribute can be weighted and assessed independently.

- **Distinctiveness**: A biometric should have features that allow high levels of discrimination in selecting any particular individual while rejecting everyone else. The larger the number of people to be distinguished, the more important this factor becomes.
- **Stability**: Age, and perhaps accident or disease, will change all biometrics over a period of time. Biometrics also may be altered by clothes or as a result of skin plasticity (e.g. smiling). A biometric should preserve enough features so that these changes will have a minimal effect on the system's ability to discriminate. Where re-enrollment can be simply or easily achieved, or where reissue over shorter durations is legally required, stability may be of less significance.
- **Scalability**: A biometric should be capable of being processed efficiently, both at acquisition time and when it is searched in a database for identification systems. Scalability issues may be less of a concern for verification-based access control systems than for large identification systems.
- **Usability**: A major selling feature in the adoption of biometrics is convenience. If a biometric is difficult or slow to use, it probably won't be adopted. Ideally, the ergonomics of the sensor will make it so simple to use that the authentication will barely be noticed. Usability is an especially important factor for people with disabilities (e.g. people who are vision or mobility impaired). This is particularly crucial in places where a biometric will be used frequently, such as for access control. Although, in some cases, mainly where a biometric is used for surveillance, this may not be relevant.
- **Inclusiveness**: An extremely high proportion of the population should be measurable, particularly for large-scale identity systems. A biometric which excludes some users causes additional complexities in managing security and has an obvious impact on usability. If the biometric is primarily for convenience rather than security, and alternatives are available, the tolerance for people who cannot use the system is increased.
- **Insensitivity**: Changes in the external environment (e.g. lighting, temperature) within reasonable boundaries should not cause system failures. Controlled indoor situations, such as airports, may be less affected than a biometric sensor used on an external door.
- **Vulnerability**: It should be difficult to create a fake prosthetic biometric (known as spoofing), or to steal and use a detached one. In some places that are highly supervised and controlled, vulnerabilities can be mitigated through policies and human monitoring.
- **Privacy**: Ideally the permission of the owner of a biometric should need to be sought before acquisition. The data should be stored encrypted.
- **Maintenance**: Sensor wear and tear, or residue build-up on the sensor surface, should be minimized. This often is achieved with non-contact sensors such as cameras, but any sensor that requires a physical touch is likely to suffer from maintenance issues.

- **Health**: Physical harm or pain should be avoided during biometric acquisition (even for those who have a medical condition, such as arthritis). Non-contact systems are much less likely to make an impact on the health of users.
- **Quality**: Obtaining a good quality sample should ideally be easy for the user. High quality samples are usually very important to ensure accurate matching results.
- **Integration**: The biometric should be capable of being used in conjunction with other authentication mechanisms, such as smart-cards or passwords
- **Cost**: The cost of the biometric system should be in proportion to the benefit. These benefits might include convenience, enhanced security, reduced cost for employing human operators or reduced cost from token loss. The cost needs to be proportional to the volume of systems that will be sold.

Biometric Attribute	(a) Laptop Sensor	(b) Passport Issuing	(c) Covert Surveillance
Distinctiveness	High	High	High
Stability	Med	High	High
Scalability	Low	High	High
Usability	High	Med	-
Inclusiveness	Med	High	High
Insensitivity	High	High	Med
Vulnerability	High	High	Med
Privacy	Med	High	High
Maintenance	High	High	Low
Health	High	High	-
Quality	High	High	High
Integration	Med	High	Low
Cost Sensitivity	High	Low	Med

Table 1.2 Biometric attributes and some example levels of importance for a) a high volume sensor used for laptop access b) a hypothetical passport issuing system using biometrics to detect fraud and c) a covert face recognition surveillance system (the covert system is non-contact so issues of usability and health are not applicable). These levels are for illustrative purposes. For real systems there may be considerable variation in the actual requirements.

1.5 Biometric Data

Biometric data is special because it is intrinsically linked to our internal concept of identity in a way that other forms of proof of identity, such as passwords and keys, are not. One of the reasons is that biometric data contains something unique and semi-permanent about the individual. Biometric data is generally represented or stored in one of three forms: *raw*, *token* or *template*.

1.5.1 Raw Data

The raw biometric information (known as the *biometric sample*) is data gathered directly from the sensor before any processing has been carried out. There is a huge range of biometric acquisition techniques - examples include camera images, infrared images, range geometry, sound recordings, chemical analysis, full motion video recordings, keystroke logs and friction, pen motion (signature) (see Fig. 1.1). Each sample mechanism has unique properties and challenges. However, there is a core set of biometrics which is more widely used in commercial applications. These common biometrics include face, fingerprint, palm, hand print, iris, speaker verification and vein matching.

1.5.2 Token Data

A token is representation of the raw data that has had some minimal amount of processing applied. For passports, the ICAO definition of the facial token to be stored on the passport chip is a cropped and scaled representation of the actual image. This is processed by the chosen matching algorithm. The reason for storing the image, rather than extracted features, is that any recognition algorithm can be used to process the 'raw' data and advances in matching are not precluded. This is known as *template interoperability*.

Another good reason for using a token is that advances in algorithms may discover new ways of extracting distinctive features from the original biometric sample. Using a token can allow seamless upgrading of algorithms.

1.5.3 Template Data

The piece of biometric data common to all biometric systems is a *template*. A template is the refined, processed and stored representation of the distinguishing characteristics of a particular individual. The template is the data that gets stored during an enrollment and which later will be used for matching.

Because of variations in the way a biometric sample is captured, two templates from the same biometric will never be identical. This is the origin of the probabilistic nature of biometrics, as the matching process can only give a decision confidence, not an absolute assurance (see Chap. 2 for more details).

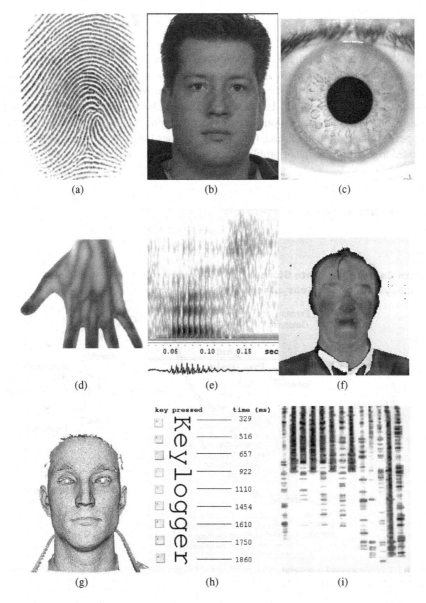

Fig. 1.1 Examples of the diversity of biometric samples: (a) fingerprint [15], (b) face [19], (c) iris, (d) vein [23] (e) voice (spectrogram) , (f) infrared face [6] (g) 3D facial geometry [10], (h) typing dynamics, and (i) DNA. Image (b) used with permission J. Phillips [20], (c) used with permission S. Phang, (h) ©2008 IEEE and image (g) used with permission P. Flynn

1.5.4 Metadata

Another source of data that is often captured in a biometric systems is *metadata*. This is data that describes the attributes of either the biometric sample (e.g. wearing glasses), the capture process (e.g. time acquired) or the demographics of the person (e.g. gender and age). The metadata is particularly important for testing and evaluation, as strong correlations between demographic characteristics and matching performance often are found. Chapter 9 examines how this information also can be used to identify groups of problem users in biometric systems.

1.6 Biometric System Overview

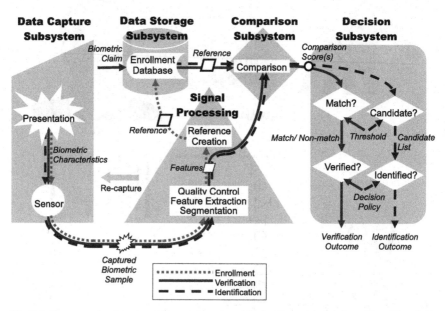

Fig. 1.2 Components of a general biometric system. Used with permission from Tony Mansfield, National Physical Laboratory, UK.

All biometric systems can be described by a general model (see Fig. 1.2). A complete biometric system includes several distinct subsystems: biometric capture, transmission and processing to enhance the biometric features, the storage of the biometric information, the biometric comparison and finally the process to decide, based on this comparison, whether it is the correct individual. Enrollment, verification and identification transactions all share related paths through the subsystems.

The first fundamental stage is *data capture*, which is the capture of the raw biometric information from a real person, which is translated into a digital sample. For many biometrics this involves a transformation from analogue to digital. Common forms of data capture equipment include digital cameras (for face recognition), optical and capacitance fingerprint scanners (fingerprint), and microphones (speaker verification). Each transformation from the original signal into the digital representation involves the potential addition of noise. Due to variance in the presentation, no two samples will ever be exactly the same.

Once the captured biometric sample is in a digital form, it is transformed using *signal processing* techniques into reference features that are used distinguish the individual. This involves processing the sample to remove noise or unnecessary background and extracting features. In face recognition it may involve finding the eyes, and in fingerprints the minutiae (i.e. ridge endings or bifurcations). *Quality assessment* is also made at this stage (see Sect. 3.3), and where the quality is poor, a re-capture of the data may be necessary. After the features have been extracted and are of sufficient quality, a reference template is created.

At enrollment, a template is created and then the *data is stored* in a database or on a device such as a smart-card. Data may also need to be protected by encryption for both security and privacy reasons.

A biometric algorithm will take the features from the stored reference template, along with the features extracted from the presentation sample, and compare them to generate a score which indicates the likelihood that both are from the same person (comparison subsystem). The output comparison score may come in a variety of forms, such as from zero to one, unbounded, or such that the closer the score is to zero, the more likely the match. This score is the fundamental building block of the analysis techniques presented in this book.

For verification, the comparison score is used to make a *decision* about accepting the person as genuine or rejecting them as an impostor. Alternatively, during identification the match is conducted against two or more enrolled people to produce a candidate list of possible genuine matches. The decision policy of whether to accept or reject them should be based on a sound understanding of the true likelihood of a mistake, and is discussed in detail in Chap. 7.

Two other components not shown in this diagram are the transport subsystem, which is the mechanism by which data is moved securely between the different subsystems, and the administration subsystem, which allows the management of the biometric system including setting system thresholds and administration of templates. Also not shown, but implied by the signal processing and the data capture subsystems, is liveness detection to prevent against the presentation of fake biometrics.

1.6.1 Negative Identification

Biometric systems usually conduct searches of a database in order to determine if an individual is enrolled, and this is known as *positive identification*. The opposite of this situation, *negative identification*, is to confirm that a user is *not* enrolled. In this case, a successful query is defined by the *absence* of any match results. A common usage is to prevent the creation of multiple identities for a single user.

Negative identification reverses the meaning of a 'false match'. For negative identification, a false match leads to a false rejection, whereas for positive identification it leads to a false acceptance. This can be illustrated by considering a passport issuing system. It is necessary to ensure that each person is issued with only one active passport, and this can be achieved by looking for duplicate matches in the list of people who already hold a passport. In this negative identification scenario, a 'false match' means that we incorrectly identify an applicant as an existing holder, and no passport is issued. However, when the passport is used at an automated immigration gate, a false match would mean that the user was in possession of another person's passport.

This also illustrates that the types of vulnerability are different for the two scenarios. For positive identification the impostor must try to look similar to the true passport owner. However, for negative identification an 'impostor' must try not to look like themself.

1.6.2 Common Biometric Processes

The matching process of a biometric can be simplified into two phases: the capture and comparison of the biometric sample, and a decision as to whether to accept or reject the input as authentic. The first part is specific to the biometric type, while handling the decision is largely independent of the actual biometric type (see Fig. 1.3).

1.6.2.1 Biometric Specific Processes

The capture of biometric data and the matching algorithm are specific to the particular type of biometric being used. Each biometric modality has specific requirements about the way the data needs to be processed. Examples of input include one-dimensional audio data, two-dimensional images and three-dimensional geometry.

The matching algorithms need to be tuned to look for the best features to distinguish individuals, while coping with changes introduced due to aging or other variations. This generally requires them to be highly optimized for the type of biometric being matched.

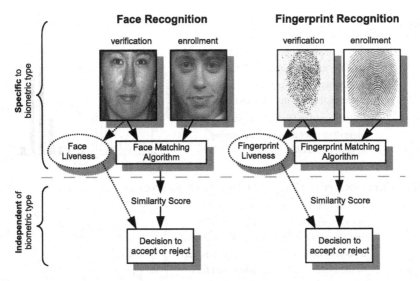

Fig. 1.3 Components of the matching process specific to the biometric type (top) and components of the matching process independent of the biometric type (bottom). Facial images from [17].

1.6.2.2 Biometric Independent Processes

The output of the biometric matching process is a similarity score. Although each different algorithm may have quite different scoring characteristics and ranges, the output represents the common attempt to assign a relative likelihood that it is a particular person and not someone else. This allows scores to be processed in the same way regardless of the biometric type or algorithms, and is the reason it is possible to compare the performance of biometric modalities and technologies despite their differences.

Match scores are either *genuine* matches, which should be high scores, or *impostor* matches, which should be lower scores. A system's performance is based on these scores, and the biometric graphs summarize this information in a useful way.

1.7 System Performance Graphs

A wide variety of graphs can be used for comparing biometric systems and representing accuracy. Many graphs are simply different ways of displaying the same data to illustrate a particular aspect of performance.

A detailed treatment of these graphs can be found in Chap. 7, while a summary of the most common types is given below.

Score Histogram

Description Plots the frequency of scores for non-matches (white) and matches (black) over the match score range. A good system will have very little overlap between the non-matches and matches.

Use Helps in understanding and visualizing algorithm operation, and for setting system thresholds.

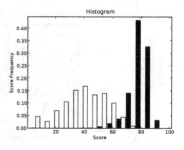

ROC Curve (Receiver Operating Characteristic Curve)

Description The ROC Curve shows the trade-off between the rate of correct verification and chance of a false match. A curve from a good system will be located near the top of the graph (high verification rate) for most false match rates. The small bars show the confidence in the accuracy of the graph.

Use For verification statistics, the ROC is commonly used to demonstrate accuracy and to compare systems.

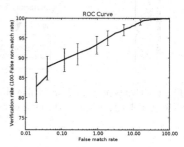

CMC Curve (Cumulative Match Characteristic Curve)

Description Displays the chance of a correct identification within the top ranked match results. A good system will start with a high identification rate for low ranks. The results in a CMC graph are highly dependent on the size of the database used for the test.

Use For identification systems, to answer questions such as "what is the chance of identifying a fraudster in the top 10 matches returned?"

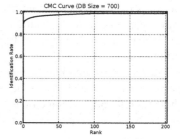

Alarm Graph

Description For systems required to generate an alarm for any matches over a given threshold, this graph shows the chance of a correct detection compared with the chance of a false alarm. A good system will have a high detection rate for low levels of false alarm. This graph is highly dependent on the size of the watchlist being used.

Use For watchlist systems, particularly surveillance, to answer the questions such as "At 1% false alarm rate, what level of detection will I have?"

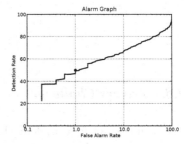

Zoo Plot

Description Displays how different users perform based on their average match score and their average non-match score. A good system will have few outliers.

Use To investigate which users, or user groups, are causing more system errors.

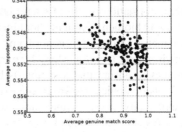

Boxplot

Description Provides a method for comparing systems at a particular false accept rate. The boxes and lines show the variation of genuine false rejection rates at this false accept rate for a given algorithm. The points represent reject rate estimates that have been classified as outliers. The lower and narrower the boxes, and the less the outlier points, the better the system.

Use To compare verification systems or different test sets.

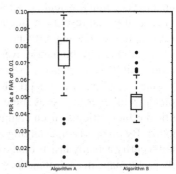

1.8 Privacy

One recurring issue around the adoption of biometrics is its potential impact on privacy. Because biometrics are bound to biological or behavioral aspects of personal identity, their use can appear to remove an element of personal control traditionally associated with other security technologies. Conversely, biometrics can help to protect privacy by combating identity fraud. When access to personal data is securely protected using biometrics as part of the authentication process, identity theft is made significantly more difficult.

1.8.1 Privacy Challenges

The central reason for considering biometric data as sensitive with respect to privacy is that your biometrics are not usually 'secret'. Furthermore, they cannot be easily changed, destroyed or declared invalid.

Covertly obtaining a biometric sample may be relatively easy depending on the biometric characteristic of interest. For example, online social network sites contain millions of images of people's faces, identity documents with facial images are used daily, fingerprints are left on almost everything we touch, and even capturing iris information covertly would be far from impossible. Indeed, the very 'public-ness' of the information is part of what makes biometrics so simple and convenient to use. A good biometric system should have some process to check that the 'non-secret' biometric information is not being forged, but also allow for normal variations in environment and use. Many biometric devices have *anti-spoofing* or *liveness* techniques to ensure that a sample is being taken from live human and not simply a prosthetic device. However, the extent and success of the system at validating the authenticity of the sample varies widely, as is discussed in Chap. 12. No systems are truly unbreakable but the aim is to make the cost of breaking the security more than any benefit that would be derived. Advances in biometric sensors, particularly in data quality and resolution, at enrollment and verification are making the task of covert capture increasing difficult.

The stability of a biometric over time is a major factor in its usability. If a biometric exhibits large changes over time, it is likely to have higher matching errors. With the exception of a fake prosthetic device, it takes major surgery to alter most biometrics in order to pass as someone else. An indirect consequence is that once a particular biometric has been compromised it is extremely difficult to replace with another. By comparison, changing a password is simple, if sometimes tedious, and good security actually insists on the continual rotation of passwords. This is particularly problematic in environments where the biometric reader or the network cannot necessarily be trusted, such as for web-based services. When the reader cannot be trusted to have read a live biometric or to have transmitted the correct results, the outcome of a biometric match cannot be trusted either.

1.8.2 Privacy Enhancing Techniques

There are a range of techniques and policies that can be used to enhance privacy.

Clever use of one way functions, or hashing, can be used to encrypt a biometric signature at the point of acquisition. Crucially, these hashes preserve the essential properties of the biometric sample, but the real sample cannot be deduced from it. The hash relies on a set of parameters that are able to be changed or destroyed, and thus can be different for each individual and for each use, allowing the 'biometric' to be revoked. This has been termed *cancelable biometrics,* as the intent is to allow the cancellation of the biometrics through this revocation process. The practicalities of such systems in commercial deployment are still to be fully evaluated as they often involve some trade-off with accuracy.

The storage of biometric data on a token such as a user-held smart-card solves privacy issues, since the users are now in possession and control of their own biometric data. If a card is capable of tamper-proof biometric capture, matching, and cryptography, the issues to do with both biometric revocation and remote validation are largely resolved. This type of advanced manufacturing is not suitable for every type of biometric product, nor is it necessary in all applications. However, it does involve techniques that can be used to address privacy concerns and enhance security.

A general privacy principle revolves around the issue of informed consent. Potential users should know how their data will be used, stored and eventually disposed of, and have the right to opt-out of the biometric authentication where feasible. Policies and procedures should be enacted to provide transparency in system operation. Such policies should address *function creep* which occurs when the system's scope is gradually extended beyond what was originally intended, for instance the cross matching against other biometric databases. The use of strong audit logs helps to enforce policy settings and should provide as much information as practical on who and what has been matched.

1.8.3 Privacy Codes

There is still significant variation in internationally accepted standards relating to privacy and biometrics. One of the first nationally recognized privacy codes specifically for biometrics came into operation in Australia in 2006. This code has three new extensions to existing national privacy principles: Protection, which deals with data storage and transmission of biometrics; Controls, which ensures, amongst other things, the informed consent of users and the right to request the removal of biometric data; and Accountability, which deals with the auditing of biometric systems and privacy impact statements.

> The Biometrics Institute Privacy Code seeks to build upon the National Privacy Principles (NPPs) in a manner that provides the community with the assurance needed to encourage informed and voluntary participation in biometric programs. Biometrics Institute members understand that only by adopting and promoting ethical practices, openness and transparency can these technologies gain widespread acceptance [2].

The European Union has been proactive in its data protection regulations (*Data Protection Directive* 95/46/EC). Its basic principles are a reduction in the processing of personal data, maintaining the highest transparency possible, and ensuring individual control of processing of personal data is as efficient as possible [8].[1] The United Kingdom has established an Information Commissoner's Office as a independent authority set up to protect personal information.[2] Recent debate relevant to biometrics has centered around the use of biometrics in a proposed national identity card and on the use of biometrics in schools.

The United States has no data protection laws [22] but there are a number of codes of conduct relating to the use of biometrics, including the International Biometric Industry Association privacy principles. These principles recommends safeguards to ensure that biometric data is not misused or released without personal consent, or the authority of law, as well suggesting the adoption of appropriate managerial and technical controls to protect the confidentiality and integrity of databases containing biometric data.[3] An active body lobbying for privacy in the use of biometrics in the U.S. is the Electronic Frontier Foundation [1]. Biometric privacy related resources can be found from the electronic privacy information center (EPIC), a public interest research center in Washington, D.C., which was established to focus public attention on emerging civil liberties issues.[4]

There is significant middle ground in the privacy debate looking at the responsible and pragmatic use of biometrics. Ignoring public concerns, or failing to take them seriously, could cause a back-lash against biometric identification technology. Therefore, it is prudent for those involved with biometrics at all levels to constructively engage with privacy advocates, and be open about unresolved issues and consider how they can be addressed.

1.9 Conclusion

In this chapter biometric systems have been introduced. The biometric matching algorithms seldom sit in isolation, so it is important to understand the context of how they fit into the wider information technology and people infrastructure. Issues relating to identity management, enrollment quality, privacy, usability, education and durability can be as important to a successful biometric installation as the technical accuracy.

Nevertheless, without the foundation of high accuracy, any system will be of limited use. The next chapter covers three specific examples of how to measure and compute biometric accuracy, and understand its impact on system usability.

---

[1] http://www.dataprivacy.ie/6aii.htm

[2] http://www.ico.gov.uk/

[3] http://www.saflink.com/resources/files/WHITEPAPER_Biometrics_and_Privacy.pdf

[4] http://epic.org/privacy/biometrics/

References

[1] Biometrics: Who's watching you? http://www.eff.org/wp/biometrics-whos-watching-you (2003)

[2] Biometrics Institute privacy code. http://www.biometricsinstitute.org/associations/4258/files/2006-07%20Biometrics%20Institute%20Privacy%20Code%20approval%20determination%20FINAL.doc (2006)

[3] Caslon analytics: Indigenous marks. http://www.caslon.com.au/indigenousmarknote1.htm (2007)

[4] Bolle, R., Connell, J., Pankanti, S., Ratha, N., Senior, A.: Guide to Biometrics. Springer-Verlag (2003)

[5] Boyer, R.S.: Automated Reasoning: Essays in Honor of Woody Bledsoe. Kluwer Academic Publishers Group (1991)

[6] Chen, X., Flynn, P.J., Bowyer, K.W.: Visible -light and infrared face recognition. In: ACM Workshop on Multimodal User Authentication (2003)

[7] Cole, S.A.: History of fingerprint pattern recognition. In: Ratha, N., Bolle, R. (eds.) Automatic Fingerprint Recognition Systems, pp. 1–25. Springer (2004)

[8] Dessimoz, D., Richiardi, J., Champod, C., Drygajlo, A.: Multimodal biometrics for identity documents. Tech. Rep. PFS 341-08.05 Version 2.0, Universite de Lausanne (2005)

[9] Doddington, G., Liggett, W., Martin, A., Przybocki, M., Reynolds, D.: Sheep, goats, lambs and wolves a statistical analysis of speaker performance in the NIST 1998 speaker recognition evaluation. In: Proceedings of ICSLP-98 (1998)

[10] J.Cook, Chandran, V., C.Fookes: 3d face recognition using log-gabor templates. In: Proceedings British Machine Vision Conference (2006)

[11] Joseph P. Campell, J.: Speaker recognition: A tutorial. In: Proceedings of IEEE, vol. 85 (1997)

[12] Kanade, T.: Computer recognition of human faces. In: Interdisciplinary Systems Research, vol. 47 (1977)

[13] Kirby, M., Sirovich, L.: Low-dimensional procedure for the characterization of human faces. In: J. Opt. Soc. Am, vol. 4, pp. 519–524 (1987)

[14] Kohonen, T.: Self-organization and Associative Memory. Springer-Verlag, Berlin (1989)

[15] Maio, D., Maltoni, D., Cappelli, R., Wayman, J.L., Jain, A.K.: FVC2000: Fingerprint verification competition. IEEE Trans. Pattern Anal. Mach. Intell. **24**(3), 402–412 (2002)

[16] Maltoni, D., Maio, D., Jain, A., Prabhakar, S.: Handbook of Fingerprint Recognition. Springer (2003)

[17] NIST: The facial recognition technology (FERET) database. http://www.itl.nist.gov/iad/humanid/feret/ (2008)

[18] Pentland, A., Turk, M.: Eigenfaces for recognition. In: Journal of Cognitive Neuroscience, vol. 3, pp. 71–86 (1991)

[19] Phillips, P.J., Moon, H., Rizvi, S.A., Rauss, P.J.: The FERET evaluation methodology for face-recognition algorithms (2000)

[20] Phillips, P.J., Wechsler, H., Huang, J., Rauss, P.: The FERET database and evaluation procedure for face recognition algorithms (1998)

[21] Pruzansky, S.: Pattern-matching procedure for automatic talker recognition. In: J. Acoust. Soc. Amer., vol. 35, pp. 354–358 (1963)

[22] Woodward, J.: Biometrics: privacy's foe or privacy's friend? Proceedings of the IEEE **85**(9), 1480–1492 (1997)

[23] Yeung, D.C.S.: Forensic and security lab. `http://www.ntu.edu.sg/sce/labs/forse/ppt/ForSe-overview.ppt` (2008)

Chapter 2
Biometric Matching Basics

Chapter 1 provided a high-level overview of the field of biometric analysis. This chapter provides three different walk-through examples to build an understanding of biometric matching "from the ground up". The aim is to demystify some of the calculation of biometric statistics and explain clearly how the different performance measures are derived and interpreted. The concepts introduced in this chapter are repeated in more detail in the relevant sections of Part II.

A biometric algorithm at its core is a comparison system, taking biometric samples as input, and producing as its output a measure of similarity. This similarity (called a matching score) is particular to an algorithm and is the fundamental output of the matching process. No matter the advances in algorithms, sensors or modalities over the coming years, the fundamentals of assessing scores introduced in this chapter will remain the same.

The goals of this chapter are to:

- Examine three different scenarios, two authentication (Sect. 2.1 and 2.2) and one identification (Sect. 2.3).
- Introduce standard biometric terms, consistent with the ISO definitions, in the context of these examples.
- Provide a step-by-step introduction to how the matching process works in both authentication and identification.
- Explain with worked examples how to derive common biometric graphs (Sect. 2.2.3).
- Look at the basics of using biometrics to detect fraud (Sect. 2.3.3).

2.1 Biometric Authentication: Example 1

To illustrate the biometric matching process and introduce matching terminology, let's consider the example of a fingerprint sensor used for computer access. Such fingerprint sensors are increasingly coming as standard components on high-end laptops (see Fig. 2.1) and are now starting to appear in other consumer goods such as mobile phones and portable storage devices. Whilst the discussions here focus on laptop fingerprints, it should be clear that the same or similar arguments apply to any biometric system regardless of type or the scale of system.

Fig. 2.1 Fingerprint sensor on a laptop. The user authenticates by running their finger over the device, which produces a fingerprint image by measuring the capacitance difference between ridges as they pass over the sensor.

2.1.1 Enrollment

Before enrolling, the user must prove they are a valid user of the laptop with permission to associate a biometric credential with their log-in information. For example, this may be done by using an existing password or dealing with a system administrator.

After the user has permission, the fingerprint matching algorithm needs to learn how to recognize the chosen finger. This *enrollment* process involves presenting the finger to the sensor (often two or more times) so that the system can record all the important and distinctive details from the fingerprint. The details captured from this process are stored as a *template* in order to allow it to recognize the finger every time it is presented in the future. Users who have difficultly using the sensor, or have poor quality fingerprints, may not be able to enroll, and this is known as a *Failure to Enroll* (FTE).

Assume that the owner of a laptop has enrolled their fingerprint. In the future, when someone tries to log-in to the laptop, there are two possibilities: it can be the correct user (the person who originally enrolled) or an incorrect user (someone else who is trying to access the laptop).

2.1.2 The Correct User

When the user wishes to access the computer, he or she places their finger on a sensor, the fingerprint sample is captured, compared against the previously enrolled fingerprint template, finally resulting in a *match score*. This is known as a *genuine match*, because the comparison is between samples from the same user. The algorithm has assessed the similarly of the captured fingerprint to the one enrolled, which is then expressed by the match score. If the similarity score is sufficiently high, the user is allowed to proceed.

If the correct user attempts to gain access to the laptop but is rejected because the similarity score is too low, there are four potential reasons why:

1. The fingerprint sample captured may be of poor quality. For example, the finger might have been placed on the sensor such that only part of the fingerprint is visible. In this case, the user will need to try again. If the user is continually rejected this is known as a *false reject*, as the legitimate user has been denied access. The *false reject rate* (FRR) is the estimate of the probability that a random, legitimate user is falsely rejected.
2. Where the fingerprint quality was so bad that no matching could take place, this is called a *failure to acquire* (FTA). There are many potential causes of poor quality. For example, the finger may have been wet, or the finger was not pressed hard enough against the sensor. Further examples can be found in Chap. 3.
3. The finger used for verification is not the same as the enrollment finger. For instance, the thumb was enrolled and the index finger is used for authentication. These are known as different *biometric characteristics*, as each finger has a unique fingerprint pattern. Since the match is between different characteristics, this is called an *impostor match* despite the fact that it is from the correct user.
4. Finally, it is possible that the fingerprint was correct and of good quality, but a weakness of the matching algorithm led to a low score.

2.1.3 The Incorrect User

In the case of an incorrect user trying to gain access to the laptop, this is known as an *impostor match*. An impostor match is not necessarily from someone deliberately trying to fool the algorithm, as it could be a person who accidentally selected the wrong log-in account. The term 'impostor' in this context merely refers to the fact that the two matches are not taken from the same biometric characteristic. If an impostor match produces a low score, the person trying to log-in will be correctly rejected, and the laptop will be secure.

If an incorrect user successfully logs in to the laptop due to a high similarity score there are three possible explanations:

1. The impostor fingerprint happens to look sufficiently similar to the one enrolled that the algorithm decides that they are highly likely to be from the same charac-

teristic. This is a *false accept* as an impostor as been allowed access. The rate at which this is estimated to occur for random users is the *false accept rate* (FAR).

2. The impostor has created an artificial or prosthetic fingerprint, either through covert means or with the co-operation of the legitimate user. This is called *spoofing,* and is one example of vulnerabilities that can be found in biometric systems. This can be countered with the use of *liveness detection* that checks to see if the fingerprint is being acquired from a real finger rather than a fake, for instance using thermal sensing or checking for a pulse (see Chap. 12). The detection of spoofing is often undertaken outside the matching algorithm.

3. Finally, it's possible that the impostor fingerprint did not look very much like the enrolled finger, but the matching algorithm has a weakness that has lead to a high score.

There are other situations that may lead to an impostor gaining unauthorized access, but are not directly related to the matching algorithm. These include the coercion of a legitimate user, the physical removal and use of someone's live biometric, and inserting fake enrollments into the system. In order to prevent the last point from occurring, it is vital that strict enrollment procedures are followed, ensuring that only legitimate users have enrollments in the system. This is a key part of identity management, and is known as ensuring a secure identity chain (see Chap. 10). *Trust can only be placed in an authentication result in relation to the level of security that has been achieved during the enrollment process.*

2.1.4 The Match Threshold

The method of deciding whether to a accept the log-in as legitimate or to reject it as an impostor depends on how high or low the match score is. This decision is made by comparing the score to a fixed value known as a *match threshold*. There are four possible match outcomes depending on whether the user is the legitimate user or an impostor, and if the match score is above or below the threshold.[1]

If the biometric sample is legitimate (i.e. a genuine match) and the score is above the threshold, the user will be accepted. However, if the match score is under the threshold, the legitimate user will be falsely rejected and may be asked to resubmit the biometric. For biometric matches with the incorrect biometric instance (i.e. an impostor match), if they are over the threshold then the biometric has been falsely accepted. However, if the match score is under the threshold, then the impostor has legitimately been denied access (see Table 2.1).

---

[1] In some algorithms the scoring is reversed such that a low score means a better match. For such algorithms, simply swap the meaning of high and low scores, and above and below the threshold.

Score	Genuine Match (legitimate sample)	Impostor Match (incorrect sample)
Score ≥ threshold	*Correctly accept*	Incorrectly accept: security breach
Score < threshold	Falsely reject: user inconvenienced	*Correctly reject*
No Score - sample quality is too poor	Failure to acquire (user must re-try)	

Table 2.1 Different options for impostor and genuine depending on the score.

2.1.5 Matching Performance

While the score determines whether the user is accepted by the laptop, it does not indicate if it is actually a genuine or impostor match. In fact, an operational algorithm never knows for certain if it is seeing an impostor (as they don't usually identify themselves), so the system must rely only on the score to make its decision.

To test the fingerprint algorithm both impostor and genuine matches must be conducted to work out the chance of an incorrect result. During testing it is known whether matches are impostor or genuine, so the scores can be appropriately labeled. The labeling of matches as impostor or genuine is known as establishing *ground truth*.

There are a number of standard graphs that are used in biometric testing. The most fundamental is the *score histogram*. The score histogram shows the frequency of scores for both genuine and impostor matches over the full range of possible scores. In other words, it represents the probability distribution of the scores. Ideally, the majority of the *genuine distribution* and the *impostor distribution* are well separated, with the genuine distribution being comprised of higher scores (see Fig. 2.2). It is important to understand the meaning of histograms as interpreting biometric evaluation results requires an appreciation of the difference between the genuine and impostor distributions.

Each genuine fingerprint presentation should appear similar to the corresponding original enrollment, although it will never appear exactly the same. Consequently, there will exist some people who have very high genuine scores and others who have relatively low genuine scores. Table 2.2 contains examples of common reasons for genuine fingerprint matches achieving low scores. For some people a particular algorithm may find recognition inherently hard because the biometric features used by the algorithm are missing, or difficult to detect. Users who fall into this category have traditionally been called *goats*. For example, our laptop user may have difficulty because they are always using hand cream, which can cause matching difficulties as some sensors are disturbed by the extra coating on the fingers. Additionally, if our user is from a demographic that has smaller than average fingers, the smaller features could cause matching problems. The formal definition of goats,

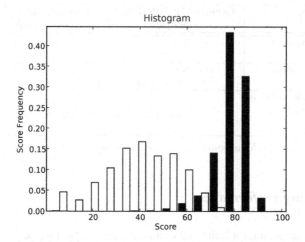

Fig. 2.2 A match score histogram for the hypothetical laptop fingerprint system. The black bars represent the genuine distribution (the range and frequency of genuine scores) and the white bars represent the impostor distribution (the range and frequency of impostor scores).

as well as other animal types and detection methods, is the subject of discussion in Chap. 8.

Low Genuine scores	High Impostor Scores
Time difference e.g. Enrolled when 12 and verifying when 24 (finger size has changed)	**Fraud (or ground truth error)** e.g. Two users who are labeled as the different people are actually the same
Poor quality (enrollment or verification) e.g. Dirty fingerprints	**Similar prominent biometric features** e.g. All users with similar fingerprint patterns receive high impostor matches
Inherent poor distinguishing features e.g. Workers with fingerprints that have been worn off through using abrasive substances	**Weak templates (lambs)**: a) Light fingerprints with lots of spurious minutiae that are unusually easy to score high against. b) A fingerprint swipe sensor that is too small leading to few minutiae per swipe.
Environmental e.g. Fingers are captured in a place with high humidity.	**Environmental** e.g. The sensor has latent prints that are left on the sensor and matches are conducted against these.

Table 2.2 Some common reasons for receiving low genuine and high impostor scores with examples from fingerprint matching. There are many others causes of high genuine and low impostors that are related to the algorithm and the inherent differences been people in the test set.

It is also expected that impostor scores will also vary across a range, and some causes for this variation can be seen in Table 2.2. One cause of note is where a weak enrollment is created that causes high impostors scores for an individual (known as *lambs*). These may be due to a lack of distinguishing features or to features that match strongly against others (in other words they do not represent strongly unique features for that person). People who get high match scores against these lambs are called *wolves*. Each person has individual genuine and impostor distributions, and these can vary significant from user to user. The individual score distributions form the basis of what is known as the *zoo plot* (see Chap. 8). The zoo plot is a technique for establishing correlations between how different users perform. It allows problems related to both the individual user data (i.e. poor enrollment quality) and the algorithm (i.e. problems with ethnic demographics) to be uncovered.

The distributions calculated rely only on the data being used and the algorithm under evaluation, and they may not reflect inherent properties of either the data or the algorithm in isolation. A test set consisting only of male office workers between the ages of 25 and 30, and using a laptop fingerprint sensor that works at 500 dpi (dots per inch) will not tell you how well the system will perform if the scanner resolution is increased to 1000 dpi[2], let alone how the system might perform on the wider population. In summary: *Change either the data or the algorithm and it will alter the calculated performance.*

2.1.6 Setting a Threshold

The chance of the fingerprint algorithm in the laptop correctly accepting or rejecting a biometric match depends directly on the threshold that has been set. When setting this threshold value, there is a trade off between falsely preventing legitimate users from logging in or making the system difficult to use, and falsely accepting impostors, making the system insecure. In order to determine the optimal value for a threshold, it is necessary to do testing, both with genuine users and impostors, to establish what score range the algorithm produces for each match type.

2.1.6.1 The Equal Error Graph

For our laptop users, it is necessary to set an appropriately secure threshold, and there are a number of graphs that can be used as aids for this purpose. They can all be generated, or at least understood, from the data presented in the score histogram. In fact, most graphs actually represent different ways of looking at the same underlying algorithm performance. On the histogram, a threshold can be picked (a point on the x-axis) and the number of impostor scores above this value is counted (these are the false matches). The proportion of these compared to the overall number of impostor matches gives the false match rate at that threshold. For the genuine scores, the proportion of scores that fall below the threshold gives the false non-match rate. When this process is repeated over a range of threshold values, a graph can be generated that shows the false match rate against the false non-match rate for

---

[2] At 1000 dpi it becomes possible to see individual sweat pores on a fingerprint, and this can be used as part of the matching process.

any threshold (see Fig. 2.3). This allows a direct reading of the trade-off between making the system easy to use (few false rejections) and highly security (few false acceptances).

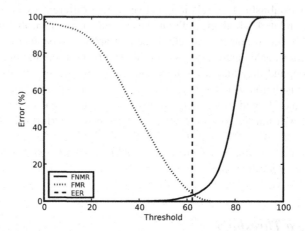

Fig. 2.3 The false match rate and false non-match rate for the hypothetical fingerprint system. In this case the equal error (where the FMR=FNMR) occurs at a threshold of 62.5 with an error of approximately 3.4%. This graph is for the same match data used in Fig. 2.2.

In some cases the threshold will be set lower to ensure that laptop users are not inconvenienced, while in other cases it may be set higher for applications requiring more stringent security. At ridiculous extremes, the threshold can be set to let any user log-in, and never detect an impostor, or to reject all users, and hence ensure that no impostor users are accepted. If no impostor testing for fake users is done on a system, it may appear to be working very well (as legitimate users are never being rejected), and yet be very vulnerable to accepting impostors. This highlights the importance of on-going testing, and the reason for ensuring a deeper understanding of how systems are operating.

2.1.6.2 The ROC Graph

The graph that is most commonly used to help select thresholds is called an ROC (Receiver Operating Characteristic) curve (Fig. 2.4). This graph takes the the false match/correct match (this is the verification rate, or 100-FNMR) rates and plots them against each other for a range of thresholds. Each point on the graph allows one to read the false match rate for a given verification rate or vice versa. Traditionally, the false match rate axis is logarithmic, as the low values are of more interest (to improve security). This allows answers to question along the lines of "if I want to

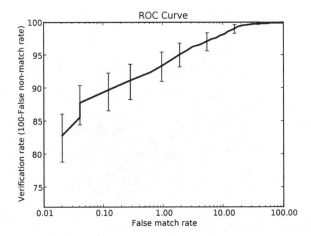

Fig. 2.4 The ROC curves for our hypothetical fingerprint system with confidence intervals. At a false match rate of 1%, the system's verification rate is 93%. The lines at regular intervals are the 95% confidence intervals, demonstrating the level of significance of the verification results. At a false match rate of 1% the confidence interval is about ±2% in the verification rate. This graph is for the same match data used in Figs. 2.2 and 2.3.

have a false match rate of 1 in 1000, what percentage of legitimate users will be able to correctly pass?"

For lower values of the false match rate, there is likely to be much less data as this corresponds to one of the extreme ends of the impostor distribution. Hence, the ROC curve for small data sets will often look quite jagged in this region, as can be seen in Fig. 2.4.

2.2 Biometric Authentication: Example 2

Following on from the example on matching fingerprints, let's now examine the situation where our laptop users are also going to be authenticated using their faces at the door to the office. This section builds matching scores and graphs by hand for a small and simple test authentication situation with 10 users.

Each user in our system has been issued a card (only given to them after an appropriate check of their identity) that encodes a unique user ID, and also includes a face recognition template encoded from the enrollment image. Images are captured and matched at the door (Fig. 2.5) and the system needs to be set up to reject impostors, and let in genuine users. To work out the best settings, we build up a score histogram, derive the ROC curve, and then determine an appropriate matching threshold.

Fig. 2.5 Card door access. A camera at the door takes a photo and compares it to the facial template stored on the card to determine whether or not to allow access.

2.2.1 Matching Data

To begin, the 10 access cards are handed out - some to the genuine users and some to people who are not the correct card holders. Each person in the test presents their card, labeled (A-J), to the door and a match score is generated depending on the similarity of the enrollment stored on the card and images collected from camera. A score of 1 means that 'it is almost certainly a different person', a score of 3 is 'unsure', and a score of 5 is 'it is almost certainly the same person'. Where a match cannot be done because of poor image quality (perhaps because there was not enough light) a 'Fail' is placed next to the image. A set of sample scores for our simple system are in Table 2.3.

2.2.2 Ground Truth

The ground truth is as follows: the first four images of Tab. 2.3 (labeled A-D) are genuine (i.e. images from the user card and the images captured from the door are from the same person) and the remainder (E-J) are impostor (i.e. card user is not the correct person).

The number of scores for each of the matches labeled as genuine and impostor is counted to give the following table:

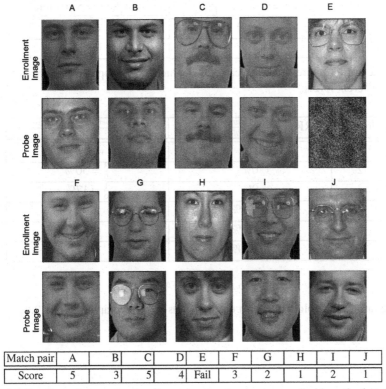

Match pair	A	B	C	D	E	F	G	H	I	J
Score	5	3	5	4	Fail	3	2	1	2	1

Table 2.3 Enrollment images stored on the card and capture images acquired from the door. Sample match scores are given between 1=bad match and 5=good match. Images used with permission from J. Phillips [2].

	Score=1	Score=2	Score=3	Score=4	Score=5	Failure
Genuine	0	0	1	1	2	0
Impostor	2	2	1	0	0	1

Table 2.4 Genuine and impostor score distribution for the match pairs in Table 2.3.

2.2.3 Calculating Error Rates and Graphs

Table 2.4 shows the number of matches that have been assigned as genuine and impostor for each score. One can see that, as expected, low scores are more likely to be impostors, and high scores more likely to be genuine. The values in this table can be turned into performance rates by dividing the cumulative number of matches by the total number of genuine matches (4) and impostor matches (6):

	Score=1	Score=2	Score=3	Score=4	Score=5	Failure
Genuine	0	0	$\frac{1}{4}$	$\frac{1}{4}$	$\frac{1}{2}$	0
Impostor	$\frac{1}{3}$	$\frac{1}{3}$	$\frac{1}{6}$	0	0	$\frac{1}{6}$

To determine a threshold, it is calculated what would happen if the threshold was set at different levels:

Threshold	FRR: $\frac{number\,genuine < threshold}{total}$	FAR: $\frac{number\,impostor \geq threshold}{total}$
1	0%	$\frac{5}{6}$ = or 83%
2	0%	$\frac{3}{6}$ = 50%
3	0%	$\frac{1}{6}$ = 17%
4	$\frac{1}{4}$=25%	0%
5	$\frac{2}{4}$=50%	0%
6	$\frac{4}{4}$=100%	0%

Failure to Acquire (FTA)= $\frac{1}{6}$= 16%

These rates can be plotted against each other as an ROC graph (see Fig. 2.6).

2.3 Biometric Identification: Example 3

The final example considers a scenario of issuing access cards. Users will get their first card free, but must pay for subsequent cards. Three people, all of whom claim to have not had an access card issued previously, arrive to collect their free card. Before issuing them with a card, the face recognition system is used to see if they have already been issued a card (this is analogous to making sure there are not duplicate identities in a passport or driver's license system). This situation is known as negative identification, as success is marked by there being no relevant matches. The person's claim of identity is ignored and the database is searched for potential biometric matches. The aim is to determine if the new person is already in the *gallery* of past users, and if so, what their real identity is. Table 2.5 shows the gallery of photos from the biometric database of people with access cards.

2.3.1 Matching Data

Assume the three people to be issued new cards are Tim, Mary and Jane. When the photo is taken for the new enrollment, a match is done against every user in the

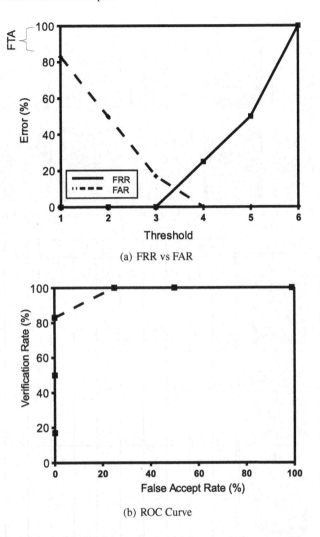

(a) FRR vs FAR

(b) ROC Curve

Fig. 2.6 (a) Match rates and (b) ROC curve for the simple example face recognition entry system.

gallery. The results of each individual match are represented by, as in the previous example, a score of 1-5, which varies from 'quite sure it is a different person' to 'almost certain it is the same person'.

Table 2.6 shows matching scores that have been assigned by comparing the new people (the probes) to each person in the gallery. With such a small database this is simple, but the larger the database, the more likely it is that there are matches that look similar to the probe, making it a challenging task to distinguish the correct person.

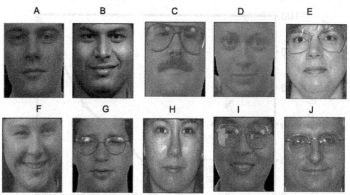

Table 2.5 Gallery images of enrolled users. Images used with permission from J. Phillips [2].

Gallery Person→	A	B	C	D	E	F	G	H	I	J
(a) Tim	2	4	1	1	1	1	1	1	1	1
(b) Jane	1	1	1	3	1	2	1	1	1	1
(c) Mary	1	1	1	1	2	1	1	3	1	1

Table 2.6 Matching results for Tim, Jane and Mary against the enrolled users. Images used with permission from J. Phillips [2].

2.3.2 Candidate List

When the results are sorted by score the top two candidate matches for each person can be shown (Table 2.7).

The ground truth for this example is Tim had card B, Jane had card E and Mary was not enrolled. These results demonstrate three identification possibilities:

	Rank 1	Rank 2
Tim	B, Score = 4	A, Score = 2
Jane	D, Score =3	F, Score =2
Mary	H, Score = 3	E, Score = 2

Table 2.7 Top two candidate matches for each new user

1. **Correct rank 1 match**: Tim was previously enrolled as person B. A *rank 1 match* has occurred, as the true gallery image had the highest score.
2. **Incorrect rank 1 match**: Jane is person F and was ranked second. The rank 1 match is person D. If a candidate list of size of two was used, a human operator might have spotted the correct person.
3. **Not in the watchlist**: Mary is not in the enrollment gallery, but has matched with a score of 3 against person H.

For our example, it was only Mary who should be issued a new free card.

2.3.3 Rank-based vs threshold-based candidate list membership

If the operator only looked at rank 1 matches, then Jane might have been mistakenly given a card. To increase the chance of identification, it is often the case that the operator will look at the top ranked 5-10 matches. The Cumulative Match Characteristic (CMC) curve allows the judgment to be made regarding how many results an operator can practically look at, versus the chance that the true person will be included in the candidate list. Note that if there is more than one image of Jane in the database, it is more likely that she will appear near the top of the rankings as there will be several chances for a high scoring match.

For some systems, sometimes known as *lights-out* matching, the operator only investigates if the system produces a match above a predefined threshold. This type of alert is known as an *alarm,* and the threshold over which alarms are generated is known as an *alarm threshold*. With an alarm threshold of 3, our current example demonstrates three different possibilities:

1. Tim would have correctly had an alarm raised.
2. Jane would have incorrectly had no alarm raised as she achieved a score of 2.
3. Mary would have correctly had no alarm raised since she was not enrolled.

Thus, at an alarm threshold of 3:

- The correct detect rate is $\frac{1}{2} = 50\%$, as there were two events (1,2) that should have triggered an alarm and only one did.
- The false alarm rate would be $\frac{0}{1} = 0\%$, as there was one event (3) that should not have triggered an alarm and it did not.

This can be repeated over a range of thresholds to build up an alarm graph so that an appropriate alarm threshold can be selected. Alarm graphs are particularly useful for surveillance analysis, and where a system will see non-enrolled individuals (known as an *open-set system*). There are a number of subtle variations in the calculation of performance for these systems, and these are covered in detail in Sect. 7.2.2.4.

2.4 Conclusion

This chapter has presented some simple illustrative examples of how biometric systems make decisions. Whilst the examples chosen used fingerprints and faces, the same techniques apply to any biometric. A more formal introduction to the generation of these statistics relating to biometrics can be found in Chap. 7.

You are encouraged to attempt the examples in this chapter using your own personal scores instead of those provided. This should allow you to work out your own personal best threshold for recognizing people. Some research from the University of Texas in 2007 [1] on the performance of people versus face recognition algorithms has shown that the more advanced algorithms actually surpass average human performance for many classes of problems.

As should be appreciated from the worked examples in this chapter, the quality of the enrollment and probe data used for matching is a vital to assessing and ensuring system performance, and the next chapter introduces the common sources of variations for different biometric modalities.

References

[1] O'Toole, A.J., Phillips, P.J., Jiang, F., Ayyad, J., Penard, N., Abdi, H.: Face recognition algorithms surpass humans matching faces over changes in illumination. In: IEEE Transactions on Pattern and Machine Intelligence, vol. 29, pp. 1642–1646 (2007)

[2] Phillips, P.J., Wechsler, H., Huang, J., Rauss, P.: The FERET database and evaluation procedure for face recognition algorithms (1998)

Chapter 3
Biometric Data

Biometric data has well defined relationships between its various data elements: people, templates, samples and matches. An understanding of the associated structure is fundamental to both building robust biometric systems and the analysis of data relationships. However, many systems that are developed do not appropriately reflect these foundations, and as a consequence are less flexible than desired. The ongoing evolution of biometric standards is also helping to enforce data quality standards, and facilitate interoperability and data exchange between different biometric systems.

Every biometric has unique properties. In addition to the wide variety of physiological differences, the acquisition process introduces many differences in sample appearance and quality. Determining the variations that lead to poor performance is vital to the analysis of any biometric system. According to the Pareto principle, it is likely that 80% of problems in a system are due to just 20% of poor quality enrollments and acquisitions. Consequently, examining issues related to biometric data is useful and informative, as it gives an appreciation for real world challenges in deploying a biometric solution.

The goals of this chapter are to:

- Explain the inherent relationships between people, templates, biometric data and matches (Sect. 3.1).
- List some published biometric standards in data interchange, applications and testing (Sect. 3.2).
- Show examples of quality variation for several commonly used biometrics (Sect. 3.3).

3.1 Storage of Biometric Data

Biometric information comes in many forms including personal biographic details, match results, acquisition and enrollment times, sensor types, raw biometric samples, errors, templates, quality information and scores. Structuring this information to allow for any type of system use and expansion is an important part of biometric system design.

3.1.1 Primary Biometric Data Elements

The relationship between the different primary data elements in a biometric system is shown in Fig 3.1. This structure holds true regardless of the biometric or whether the system is used for identification or verification. Databases conforming to this Primary Biometric Data Element (PBDE) structure will have increased flexibility.

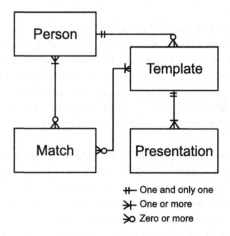

Fig. 3.1 Primary Biometric Data Elements (PBDE) structural relationship.

- **Person**: The top level entity is the 'Person', from which all other data is derived. Each person may have associated with them biographic data that does not change, such as sex, date of birth and ethnicity. A person entity can be associated with zero or more match records and zero or more templates. For example, they may have templates from several different characteristics or templates acquired at different times.
- **Template**: Each person may have one or more biometric templates. These templates contain both enrollment, and potentially verification, information.[1] Associated with each template can be a variety of data including a quality measure, the date of capture and information specific to the biometric or environment such as the ambient lighting or if a person is wearing glasses. A template belongs to one

---

[1] The term template is used in a slightly different manner than from the ISO definition since it may refer to a "template" generated from a verification sample as well as from an enrollment sample.

and only one person, but may be constructed from multiple biometric samples. A match is the comparison of two templates.

- **Presentation**: A template may be created from the sample data aquired during one or more biometric presentations. When multiple samples are used as part of the template this is usually created using several presentations of a biometric over a short time period during enrollment. A system may also hold biometric presentations from different modalities (e.g. finger and iris). Generally, one particular presentation only contributes to a single template.
- **Match**: A match record contains the score result from a comparison of two template records. It can be either a genuine match, where the templates are from the same person and instance, or an impostor match, where they are known to be from different people. The record also needs to record matching errors (such as a failure to acquire). A match may belong to either one person, for a genuine match, or two people, for an impostor match.

3.1.2 Transactions

An authentication using biometrics may be comprised of a series of matching attempts, known as a transaction, rather than a single match. In other words, the decision to accept or reject the identity claim may involve multiple matches. One example is a system that allows a second (or even third) attempt to verify if the first attempt fails. Alternatively, there may be sequential matches that take place during authentication, such as in a speaker verification system that asks several questions for verification. Finally, some systems are multimodal, and these systems use more than one biometric, and hence more than one match must be conducted (see Chap. 4). A transaction record must group all these matches into a single logical unit along with a single verification decision (i.e. accept or reject). In the entity relation structure (Fig. 3.1) this can be accomplished by the use of a transaction identifier for each match record.

3.1.3 Errors and Quality

For each match or enrollment there are a variety of possible errors that may occur. Most biometric errors come in the form of a Failure To Acquire (FTA) or a Failure To Enroll (FTE). Recording the reasons for failure, often due to quality factors, can assist in determining how to improve overall system performance. Note that recording failures may result in needing to create a place holder for a "failed" template, since the template was not created. Recording the quality of the sample in the template record is also recommended since this allows for modification of system thresholds to fine tune matching results for marginal quality cases, and is particularly useful when a new or updated algorithm is introduced.

The calculation of a quality score is the subject of much research. As shown in Sect. 3.3, there are many factors that can influence quality, and these are specific to the biometric being used. In general, a quality score should reflect a sample's expected matching performance. A unified quality score may involve the combination of many individual aspects of the biometric sample (such as for the ICAO face quality), or in many cases the matching engines will produce quality score as part of the matching process. Quality scores are discussed further in Sect. 8.1.3.

3.1.4 Upgrades

Biometric algorithms are constantly evolving and the design of databases to store biometric data should consider the impact of a requirement to upgrade or change engines. A new engine may have a different template structure from older engines. In this case, if the original template data has been discarded all users will need to be re-enrolled, which is a potentially costly process. Where match results are labeled with the engine and version, any analysis or audit can determine which algorithm was used for the creation.

3.1.5 Data Security and Integrity

Data security and integrity are vital to ensure trust in large scale biometric systems. Recent large scale identity thefts and losses have illustrated how important it is to consider the protection of the data. Many of the security techniques suggested are best practice for any information technology based system [1] (see Chap. 12). Physical and logical controls on systems storing biometric data can be used to prevent system breaches. Examples of these data security techniques include:

- **Data De-identification**: When possible, it is recommended that name and contact information is removed and held in a physically separate storage system than the biometric data. One advantage of this is that it reduces the impact of a system breach. It should be considered mandatory that before data is transferred over an untrusted link all records are de-identified.
- **Data preservation and signing**: Information should not be destroyed, and where possible the originals should be preserved. This is particularly important if the data may be used as evidence at a later time. Data should also be securely signed to detect any alteration.
- **Data Encryption**: Encryption is a process by which data is rendered difficult to read by unauthorized parties. Encryption targets may include any sensitive records, such as the sample data and the templates. The encryption process must include the proper management of encryption keys through a certificate authority, or similar mechanism. Encryption can also act as a barrier to the addition of identification searches when the requirements were only for verification, also

known as function creep[2], since it is more difficult to undertake a one-many search on a database where each template has been encrypted separately.

Cryptographic Approaches

Encryption is undertaken using cryptographic algorithms. These algorithms fall into two classes: symmetric and public key.

- Symmetric algorithms are where the same key is used for both encryption and decryption. This key must hence be tightly protected and cannot be shared with untrusted parties.
- Public key algorithms use two keys: a public encryption key and a private decryption key. The public encryption key may be shared, because if the message is intercepted it cannot be read without the private key. This allows much tighter controls around data security since the points at which encryption is required, such as biometric readers, do not need to store sensitive keys to securely transmit data.

Other useful cryptographic techniques include digital signatures, that can be used to prove the authenticity of the message as originating from a particular sender, and one way hashing functions that can provide assurance that the message has not been tampered with. The use of cryptographic approaches should be applied with appropriate specialized expertize, as it is easy to provide an apparently secure solution that has significant vulnerabilities, as several large companies have learned the expensive way.

- **Audit controls**: Strong audit controls can be implemented to ensure that any operations undertaken are recorded and traceable to a particular authorized individual. Audit logs should be sent securely to an unalterable data repository and digitally signed on a regular basis. They should also be recorded in an easily read non-proprietary format.
- **Data removal**: Data should be removed when it no longer needs to be held for archival or operational purposes. Possible reasons for data removal include expiry after a particular time period, due to an event such as the removal of an enrolled person, or as a result of information redundancy.

---

[2] Function creep is the addition non-intended functionality after a project is complete. For example, identification facilities when the original requirements were only for verification.

Storing Raw Data

The raw sample data used for the creation of templates is often deleted to protect privacy and reduce storage requirements. However, without the source data changing or upgrading the matching algorithm will be more difficult, or even impossible in some cases. Also, forensic use of the data as evidence may not be possible without the original biometric sample. A compromise is to store the original data using protected write-only or offline storage facilities, such that the data is archived but cannot be accessed on demand.

3.2 Standards

There has been a significant amount of work in the development of standards for the interoperable storage, transport and use of biometric data [13]. The standards come in three forms: those dealing with generic biometric services and data, the interoperability of different biometric characteristics and the standards on the testing of biometric systems. The standards are constantly developing, with some still in a draft stage and others, such as the BioAPI 2.0 specification, being quite mature. Standards frequently start as national standards (e.g. ANSI, American National Standards Institute) and progress to become international (ISO, International Standards Organization). The ISO Joint Technical Committee One (JTC1) contains the relavent subcommittee SC37 for biometrics, other subcommittees of interest include SC17, for cards and personal identification, and SC27 for IT security techniques. The following sections provide a snapshot of a selection of some relevant standards.[3]

3.2.1 Formats for Data Interchange

- **Finger Pattern**: An interchange format for the exchange of pattern-based fingerprint recognition data is defined by this standard. It includes both the conversion of a raw fingerprint image to a cropped and down-sampled finger pattern, and the representation of the finger pattern image. *M1 ANSI INCITS 377-2004*
- **Finger Minutiae**: The representation of fingerprint information using minutiae (ridge endings and bifurcations) is defined by this standard. It includes the placement of the minutiae, a record format, and optional extensions for ridge count and core/delta information. *M1 ANSI INCITS 378-2004*

---

[3] Where only ANSI standard numbers are given the standards also exist as parts standards under ISO/IEC SC37

- **Finger Image**: Image-based fingerprint and palm print recognition data exchange formats are defined by this standard. This standard is intended for those identification and verification applications that require the use of raw or processed image data. *M1 ANSI INCITS 381-2004*
- **Iris**: This standard defines iris attributes, a record format, sample records and conformance criteria. Two alternative formats for iris image data are described, one based on a Cartesian coordinate system and the other on a polar coordinate system. *M1 ANSI INCITS 379-2004*
- **Face Recognition**: Photographic (environment, subject pose, focus, etc.) properties, digital image attributes and a face interchange format for relevant applications are defined in this standard. This includes both human examination and computer automated face recognition. *M1 ANSI INCITS 385-2004*
- **Signature/Sign Data**: A Signature/Sign Data interchange format is defined containing definitions of raw, time-series based signature/sign sample data and signature/sign feature data as well as a data record format. *M1 ANSI INCITS 395-2005*
- **Hand Geometry**: A hand geometry data interoperable interchange format is defined. *M1 ANSI INCITS 396-2005*
- **Speaker Recognition**: Draft standard to define an interoperable data format for speaker verification systems. *ISO/IEC JTC 1/SC 37 N 1973*

3.2.2 General Standards

- **The BioAPI 2.0 Specification**: The BioAPI provides a high-level generic application programming interface and service provider interface for any biometric technology. It is designed to allow 'seamless' connection of different biometric sensing equipment and algorithms. An increasing number of manufactures have products that are compliant. *M1 ANSI INCITS 358-2002, ISO/IEC FDIS 24708*
- **Common Biometric Exchange Formats Framework (CBEFF)**: The Common Biometric Exchange Formats Framework (CBEFF) is a standard to allow the generic holding and transmission of biometric information along with associated metadata in a standard form. It does not attempt to specify the format of the biometric template, and acts primarily as a container. It is used as the transport encapsulation for BioAPI. An XML version of CBEFF also exists. *M1 ANSI INCITS 398-2005*
- **Multimodal and other multi-biometric fusion**: Describes standardization to support multi-biometric systems using various multi-biometric fusion techniques. *ISO/IEC TR 24722:2007*
- **Biometrics Tutorial**: Describes the main biometric technologies and applicable international standards for biometrics. *ISO/IEC TR 24741:2007*
- **Jurisdictional and societal considerations for commercial applications**: This standard looks at issues relating to the introduction of biometrics including acceptance, eduction and privacy. *ISO/IEC DTR 24714-1*

3.2.3 Applications Interoperability and Data Interchange

- **Transportation Workers (Interoperability and Data Interchange – Biometrics-Based Verification and Identification of Transportation Workers)**: This standard defines standards for applications where tokens are used for access control and identification of employees. It is intended for use in the transportation industry and other industries where identification and verification of employees is necessary. *M1 ANSI INCITS 383-2004*
- **Physical access control for employees at airports**: Support of token-based biometric identification and verification of employees for physical access within an airport. *ISO/IEC 24713-2:2008*
- **Border Management (Interoperability, Data Interchange and Data Integrity of Biometric-Based Personal Identification for Border Management)**: Border management applications using biometrics to authenticate the identity of non-citizens as they enter, stay in, and leave the United States. *M1 ANSI INCITS 394-2004*
- **Defense Implementations (Interoperability and Data Interchange – DoD Implementations)**: Military biometric application profile settings are defined for processing and storing biometric data on enemy prisoners, detainees, internees, and persons of interest with respect to national security. *M1 ANSI INCITS 421-2006*
- **Commercial Access Control (Application Profile for Commercial Biometric Physical Access Control)**: This standard defines standards in applications that use biometrics to authenticate the identity of users requesting access to a facility. It establishes minimum conformity requirements for the biometric parts of such systems. *M1 ANSI INCITS 422-2006*
- **Financial Industry (Security framework for Biometrics in Financial services):** Describes a security framework for using biometrics for authentication of individuals in financial services. Describes the architectures for implementation, specifies the minimum security requirements, and provides control objectives and recommendations. *ISO 19092:2008*
- **ANSI X9.84**: Describes the security features needed to implement biometric verification for financial services. It focues on the integrity, authentication and confidentiality of biometric transactions. Requirements for enrollment, verification, storage, transmission, and termination procedures are documented,

3.2.4 Biometric Testing Standards

- **Biometric Performance Testing and Reporting Part 1 - Principles Framework**: This standard specifies a common set of methodologies and procedures to be followed for conducting technical performance testing and evaluations. *M1 ANSI INCITS 409.1-2005, ISO/IEC 19795-1*

- **Biometric Performance Testing and Reporting Part 2 - Technology Testing Methodology**: Procedures for conducting offline tests of the performance of biometric technologies. *M1 ANSI INCITS 409.2-2005*
- **Biometric Performance Testing and Reporting Part 3 - Scenario Testing Methodologies**: Requirements for scenario-based biometric testing and reporting. *M1 ANSI INCITS 409.3-2005*
- **Biometric Performance Testing and Reporting Part 4 - Operational Testing Methodologies**: Requirements for operational-based biometric testing and reporting. *M1 ANSI INCITS 409.4-2006*

3.3 Biometric Data Examples

Perhaps one of the best ways to appreciate a particular biometric is to look at the marginal or boundary cases for sample quality. Obtaining high sample quality at both enrollment and verification is vital for ensuring good performance in operation. The types of poor quality samples tend to fall into the following broad, overlapping categories:

- **Distortions**: Elastic distortions are non-linear distortions (e.g. stretched skin, dilated pupil), while inelastic distortions result from the distance, translation and rotation of objects relative to the sensing equipment.
- **Occlusions**: Occlusion, where part of the biometric cannot be sensed due to obstruction, can be be caused by body parts, shadows, clothing, etc.
- **Medical**: Some medical conditions result in a reduced ability to read the biometric sample. This may be due to behavioral reasons (e.g. blindness leading to difficulties operating equipment), a degradation of the biometric characteristic (e.g. a sore throat altering one's voice biometric) or the complete loss of a biometric (e.g. a missing finger).
- **Antiquity**: The process of aging can cause an enrolled biometric to differ from the verification sample, varying in degrees depending on the time since enrollment.
- **Clothing**: Clothing such as hats, glasses and contact lens can obscure or introduce sample artifacts.
- **Environment**: The environment can affect acquisition of a biometric through the introduction of noise and additional artifacts.
- **Ergonomics**: The position and ergonomics of a sensor can directly affect the quality of samples through increased difficulty of the user in presenting a good biometric sample. Examples of this include fixed height iris systems where tall or short people have difficulty using the system, or fingerprint sensors that do not have finger guides or have small sensing areas. Often poor ergonomics will lead to increased system failures through distortions or occlusions.

The following sections look at a number of common biometrics and give examples of poor quality biometric samples. Not all of the listed factors will affect a given system since it is dependent on the particular algorithm and sensor in use.

3.3.1 Fingerprint

Each of the intricate patterns on your fingers is unique. In fact, even for identical twins the corresponding fingerprints will be different. Fingerprints have traditionally been used for identification purposes by law enforcement agencies. This is typically done by matching latent fingerprints found at crime scenes against large fingerprint databases. For example, the FBI maintains a database that contains hundreds of millions of prints.

In recent years, fingerprints have been adopted for use in the field of biometric authentication (see Sect. 1.2.1). In this case, special scanners are used to capture the fingerprint. There are many mechanisms for acquiring a fingerprint image. For example, many laptops now come with a thermal or capacitance fingerprint scanner. This derives fingerprint information by reading the ridges as a person sweeps their fingerprint over the top of a silicon sensor. Technologies used for fingerprint recognition include optical sensors (essentially taking a photo of the finger), solid-state sensors that use a silicon chip to measure capacitance, thermal, electric field or piezoelectric differences on the fingerprint, and ultrasonic readers that measure ridge position using acoustic reflections.

Fingerprints have a number of structures that can be used for identification [11]:

- **Holistic ridge flow**: The overall pattern of ridges and valleys of a fingerprint can be used to classify it into one of several common categories. The most common categories are whorl, arch, left and right loop, and tented arch. Automated matching techniques tend not to rely on these classes for recognition as they are not very distinctive and are easily misclassified.
- **Minutiae**: Minutiae are locations where fingerprint ridges split or terminate. Much of a fingerprint's individuality is captured by the distribution of its minutiae points. Each minutia has a well-defined position and orientation, and comparing the relative minutiae locations between two prints forms the basis of most traditional fingerprint matching algorithms.
- **Other characteristics**: Some features, such as wrinkles, creases and warts, may also be useful for identification, however they may be transitory so should not be relied upon.
- **Pores**: With sufficient magnification and resolution, sweat pores can be seen on fingerprint ridges. The pattern produced by these pores is highly distinctive for each fingerprint.

The quality of a fingerprint image is closely related to its suitability for matching. In other words, matching poor quality images will lead to unreliable results. The following is a list of some factors that impact the quality of a fingerprint image:

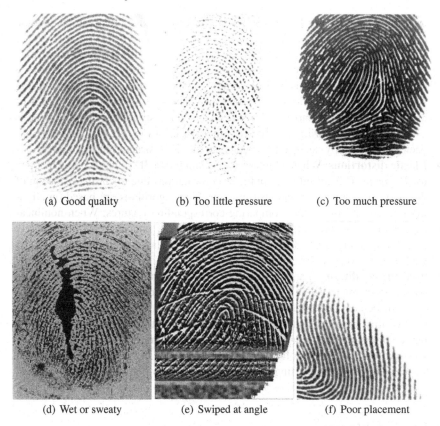

(a) Good quality   (b) Too little pressure   (c) Too much pressure

(d) Wet or sweaty   (e) Swiped at angle   (f) Poor placement

Fig. 3.2 Examples of fingerprint quality variation. (a) **Good quality**: Minutiae clearly present and ridges are well defined. (b) **Too little pressure**: Many minutiae are missing or hard to distinguish. (c) **Too much pressure**: Ridges definition is poor leading to heavy distortion and overlapping ridges. (d) **Finger wet or sweaty**: Features are obscured by moisture. (e) **Finger swiped at angle** (line-sensor): Features are distorted by the scanning process. (f) **Poor placement**: The fingerprint is highly rotated and off-center. Images (a), (b), (c), (e), and (f) are from the FVC competitions, ©Springer-Verlag.

- **Inconsistent and unreproducible contact**: Every time a finger is pressed against a surface, it is applied with a certain amount of pressure at a particular angle. The actual amount of pressure used and the contact angle will vary from time to time, resulting in a different (and incomplete) portion of the print being captured. This is especially problematic for fingerprint scanners with a small scanning surface. In Fig. 3.2 (f), the finger is highly rotated.
- **Noise**: Even under ideal conditions, noise will be present to some degree in all fingerprint images. This is an inevitable consequence of taking discrete measurements of the physical environment. More commonly, wet or dirty fingers can lead to noisy images. Furthermore, moisturizer or other hand creams can lead

to residue being left of the sensor. Figure 3.2 (d) is noisy due to smudging and moisture.

- **Incomplete ridge structure**: There are several reasons why the entire ridge structure of a fingerprint may not be captured. If the skin is dry, sweaty, diseased or injured, some parts of the ridges may not make contact with the capturing surface, while some valleys may touch the surface. It is important for a verification system to be robust against incomplete ridge structures, and this issue is usually addressed during the prepossessing stage. In Fig. 3.2 (b), the finger has been pressed lightly against the scanner, so the ridge structure is incomplete.

- **Elastic distortions**: When a fingerprint is captured, a 3D finger is being mapped to 2D image. This introduces nonlinear deformations due to elastic distortion of the skin. This is troublesome as most matching algorithms are based on aligning fingerprint images and comparing corresponding features. When nonlinear deformations are present, a rigid alignment will be unable to align accurately all corresponding areas of the prints. Figure 3.2 (c) is a fingerprint image that is distorted due to excessive pressure.

- **Medical conditions**: Large changes in weight can affect the quality of some biometric characteristics. Some users may have medical conditions that make finger placement difficult, such as arthritis or Parkinson's, and this can be lead to poor quality captures or a complete inability to use the sensor. There are some rare disorders that result in very poor fingerprint definition.

- **Occupation**: In some professions where there is a high use of abrasive substances, such as building, fingerprints can be worn away or significantly scarred.

3.3.2 Facial Image

Two dimensional facial recognition is the use of information extracted from an image for enrollment and identification. It is self-evident from the ability to recognize people known to us in photos, regardless of wide variation in age and quality, that faces have some features that are distinctive and stable. However, there is evidence to suggest that people may not be as good as they expect in the recognition of non-familiar faces [14].

Face recognition systems do not usually use ratios of distances between facial landmarks, such as the inter-eye distance or length of the nose, as these are not particularly distinctive. Most recognition algorithms rely on pattern recognition using statistical learning techniques calibrated using large sets of data [22]. Significant advances in accuracy have come about over the previous years, particularly through the use of higher resolution images and algorithms that are more robust to environmental changes.

The stable and distinctive information contained in the face is focused in regions of the face that are unlikely to change - these tend to be around the central features of the eyes, nose and mouth. Many parts of the head tend to be unreliable in terms of visibility, as they may be covered by hair, hats or otherwise obscured due to rotation. Face recognition can be used with existing photos that are of poor quality or low resolution, or with images taken from variety of different camera types, so there tends to be a particularly wide range of image qualities as compared to other biometrics.

(a) Good quality   (b) Poor lighting   (c) Expression

(d) Glare on glasses   (e) Pose   (f) Low Resolution

Fig. 3.3 Facial images, examples of quality variation: (a) **Good quality**: Even lighting, neutral expression, good focus. (b) **Poor lighting**: Heavy shadowing from on the right side of the face. (c) **Expression**: Mouth smiling and eyes closed. (d) **Glare on glasses**: Lighting glare obscures left eye. (e) **Pose**: Head rotated slightly to the left (f) **Low resolution**: 22 pixels from eye to eye. Images (a)-(e) used with permission J. Phillips [15]

- **Antiquity**: The duration between the time the enrollment image was taken and the verification image was obtained can affect performance. This is because the structure of the face continues to develop, particularly during adolescence, and aging changes our skin elasticity and the surface suffers environmental damage. Hence, the effect of the antiquity on a facial image depends on both the time separation and on the age of the person enrolled. It has been observed that the older a person is, the more stable their facial characteristics for a given duration are, and the better scores that are achieved due to an increase of surface "features".

- **Pose** (inelastic distortion): The angle of the face relative to the camera introduces 3D variation in the relative position of facial features. This rotation also often causes some occlusion in facial features particularly around the nose and eye area. Ideally, the face is close to straight-on (e.g. ± 5 degrees) to the camera. In some modern systems rotated images are synthetically generated during enrollment, which improves recognition accuracy when the verification image is also rotated. In Fig. 3.3 (e) the head is rotated to the left and in Fig. 3.3 (f) the head is both highly rotated and tilted.

- **Expression** (elastic distortion): The face has a large number of muscles that control the wide range of expressions that are used for human communication. A specific methodology called the Facial Action Coding System (FACS) [5] can be used to define human facial expressions. In this system there are 32 major facial variations. Each of these expressions causes an elastic distortion of the face, and the more extreme the difference in expression between the enrollment and verification images, the more difficult the matching process. Most systems recommend that the expression should be neutral at enrollment time (with no teeth shown) since this expression is likely to be common and more easily reproducible during verification. Figure 3.3 (c) has an extreme expression, with both the eyes closed and the mouth in an open smile, and Fig. 3.3 (e) has a smile showing teeth.

- **Inner Features**: The parts of the face which are most stable are commonly used as the primary recognition information. Generally, this is the inner features bounded by the eyes, nose and mouth. When these features are obscured recognition becomes significantly more difficult. For this reason most systems recommend that both eyes should be open and clearly visible, and the mouth closed with a neutral expression.

- **Outer Features**: Despite being less stable in terms of visibility, the outer features can be particularly useful for distinguishing between similar looking individuals. For example, the ear lobes and chin shape are useful and distinctive identifiers. Furthermore, when not covered by hair, even creases on the forehead or around the eyes can be used. This data is particularly useful in situations where the images are of poor quality, such as for surveillance data.

- **Image Source**: A primary measurement of source quality for the facial image is the number of pixels between the eyes, also known as the inter-eye distance. The larger this distance is in pixels the more information there is available for recognition. In particular, some face recognition systems use the skin texture given sufficient resolution. Facial features should be sharp (in focus) and have appropriate tonal information (gray levels). In most systems color information is not used since it is highly variable dependent on camera settings and environment. However, it may be useful for human recognition. The images in Fig. 3.3 all have approximately 80 pixels inter-eye distance, except for (f) which has only 22 pixels.

- **Compression Effects**: Facial images are often stored and transmitted in a compressed form. The compression process is usually lossy[4], e.g. JPEG, which creates image artifacts that can affect recognition. Where compression is used it should be set to give the best image quality practical for the storage available. Also, it should be kept in mind that if images are altered and re-compressed the information loss is incremental.
- **Lighting and Glare**: Performance can be quite sensitive to lighting conditions, particularly when there is significant variation in illumination between enrollment and verification images. The effects of uncontrolled environment lighting on the human face manifests as shadows and luminance gradients across the face. Where a single strong light source is used to create an even light, this can create significant glare on forehead or glasses, cause washout of images due to the contrast between the background and the face, or if the user is not positioned exactly in front of the light it will cause shadows. In Fig. 3.3 (b) there is strong lighting coming from the top right causing shadows on the nose and chin. In Fig. 3.3 (d) the glasses have glare reflecting from the strong light source. The best light sources are diffuse and set at an intensity that causes even lighting with no glare.
- **Glasses**: Generally clear glasses with light rims will not cause significant recognition issues. However, glasses do cause problems when they obscure the inner features by glare (see Fig. 3.3 (d)), thick rims, or tinted lenses such as sunglasses.
- **Cosmetics**: Cosmetics are used to deliberately alter the appearance. For instance, makeup may be used to make the skin look smoother, creating the appearance of higher cheek bones, or highlight the eyes or mouth. All of these actions have an effect on the facial appearance and can impact matching performance, particularly where the enrollment and verification images differ in the level or style of makeup. In large population samples, women are often less distinctive than men since they attempt to present a more uniform appearance.
- **Weight change**: Significant weight change can alter the appearance of the face, although certain facial features around the nose and eyes are less affected.
- **Natural Variants**: There are a number of demographics factors that may affect performance. These factors include ethnicity, age, beards, caps, glasses and medical conditions such as eye-patches.[5] In many cases these challenges can be addressed procedurally, such as by asking for the removal of sun-glasses and hats (see Chap. 9).

3.3.3 Iris

Iris systems use the random pattern of filaments on the front of a person's eye that regulate the size of the pupil (the iris) for identification. It has been shown that the

---

[4] The term 'lossy' refers to data compression techniques that result in a loss of information, not just file size. When lossy compression is used, it is only possible to recover an approximation of the original data.

[5] In some cases religious grounds may prevent revealing a face fully in public.

iris pattern on each eye is a highly distinctive and stable biometric characteristic. The patterns are complex and consist of a large variety of features including collagenous fibers, crypts, color, rifts and coronas. The iris pattern is set prior to birth where the iris muscle goes through folding and then de-generation [8]. After the first to second year after birth, it varies little except due to eye disease. Since the patterns are so stable, it is possible to apply specialized matching techniques to produces highly accurate results [20].

Traditionally iris cameras are set to capture an image of the iris in the near in-frared range, since it is at this wavelength that the iris structure is most apparent. Registration and alignment of the iris images previously relied on substantial user cooperation (e.g. head alignment), however recent advances now make usage much simpler. Iris systems are also starting to be combined with face recognition systems to enhance accuracy, particularly for surveillance applications. Since these images are taken under less controlled conditions it has been necessary to handle images with lower quality.

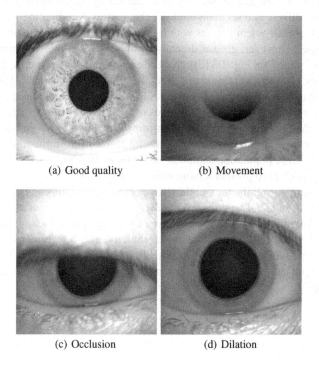

(a) Good quality   (b) Movement

(c) Occlusion   (d) Dilation

Fig. 3.4 Iris images, examples of quality variation (a) **Good quality**: fibrous structure clearly visible eye centered. (b) **Movement**: blur caused by movement during capture. (c) **Occlusion**: partially closed eye obscures some of the iris, glare on iris. (d) **Dilation**: filaments distorted and compressed to edge.

- **Contact Lens**: Contact lenses come in many forms, and not all are optically clear. Since the contact lens sits in front of the iris it will distort, to some degree, the visible filaments. If a person enrolls with a contact lens and then verifies without one this may cause performance difficulties.
- **Eye Rotation**: If the eye is not looking directly at the camera when the image is taken it may rotate about an axis. This may cause some of the filaments in the eye to be distorted.
- **Dilation**: Pupil dilation occurs when the iris muscle contracts, leading to the iris filaments being compressed. Pupil dilation can occur due to lighting extremes, arousal or drugs. Figure 3.4 (d) shows a highly dilated pupil.
- **Occlusion**: Some or all of the iris may be occluded by either the eye lid blinking, the person squinting or glasses getting in the way. In Fig. 3.4 (c) the eye is partially closed.
- **Movement**: The eye is rarely completely still due to what is known as saccades. These are fast motions in the human eye with a peak motion up to 1000 degrees per second. The movement ensures that the blind spot at the center of our vision is not noticed and allows greater visual acuity. However, it can make imaging the iris more difficult. Certain medical conditions also make these saccades more frequent. Movement can also be introduced due to normal head motion. In Fig. 3.4 (b) an iris is captured in motion.
- **Environment**: Where the iris is captured in a surveillance situation the lighting environment may cause over or under exposure of the image, making distinguishing the iris filaments difficult.
- **Eyelashes**: Long eyelashes can obscure part of the iris, causing sections of the iris to be unreadable.
- **Medical conditions**: Common medical conditions that can cause problems with iris recognition through distorting the iris include cataracts and glaucoma.
- **Glasses**: When a person is wearing glasses this can affect the optical properties of reading through the lens, particularly if the lens is tinted or has a gradient power. Glasses also collect dust and scratches which can obscure parts of the iris.
- **Glare**: Glare from lighting or environment reflection can obscure part of the iris.
- **Natural Variants**: There is a wide range of different iris variants. For example, some racial subgroups have very dark eyes, and subsequently have little visible iris structure.
- **Height**: Some iris cameras are fixed in position and require the head to be placed carefully within the capture zone. People who are shorter or taller than average, or in a wheelchair, may have difficulty positioning such that they have correct alignment.

3.3.4 Speech

Speaker verification is the use of the distinctive patterns of a person's speech for recognition. Vocal characteristics are based on both the physical aspects of the vocal chords and the episodic nature of the local accent. One of its primary uses is for the verification of telephone transactions. As speaker verification is behavioral as well as physiological, there are two types of authentication. Text-dependent recognition [21] relies on the same word or words to be spoken as were enrolled, and text-independent [2] recognition which attempts to identify a speaker regardless of what they are saying. Many complex biological factors go into the production of speech including the movement of tongue, lips, and larynx, and the relative sizes of the nasal and oral cavities. In addition, speech accent is affected by both regional and societal factors.

- **Stress**: The fundamental frequency of voice can be significantly elevated under stress conditions [17]. This is seen in raised pitch and a change in speaking cadence. Some systems have sought to use this as a simple lie detection mechanism, however since the reaction to stress varies greatly it is a rough guide at best. In Fig. 3.5 (b) the stress can be seen in higher frequency components.
- **Colds**: Colds which affect the nasal passage or the throat will have some impact on the quality of vocal data depending on severity. Where the vocal characteristics are dramatically changed it makes recognition from a good quality enrollment almost impossible. Figure 3.5 (c) shows the effect of a blocked nose on frequency response.
- **Background noise**: Most speaker systems do not operate in an acoustically isolated environment, and background noise is always likely to be present to some degree. Where the volume levels of the background are significant, the vocal frequencies can become obscured. Noises that operate in the same spectrum as the human voice will cause the worst distortion. Figure 3.5 (d) demonstrates the masking effect of a loud background noise.
- **Mobile Phones and VoIP (Voice over IP)**: Mobile phones and calls made through the Internet are highly compressed to transmit vocal data efficiently. The compression codecs causes artifacts in the vocal signal that can reduce recognition performance. In addition, drop outs caused by transmission delays or blockages also create artifacts. The effect of a poor quality mobile phone call is illustrated in Fig. 3.5 (e).
- **Channel Mixing**: When a person enrolls on one type of device (e.g. a fixed line phone) and then verifies on a different line type (e.g. a cellular phone) this is called mixing channels. Because of the different characteristics of the channels, the frequency information can be quite different. For this reason some systems require separate enrollments for each channel.
- **Speaker Phones**: Speaker phones change the audio qualities of the voice and are more likely to be effected by background noise.
- **Text recognition**: Text-dependent recognition is concerned mainly with distinguishing one speaker from another as opposed to ensuring that the enrolled word or words are actually spoken. If a similar sounding, but different, word is spoken it may still match successfully. To address this issue it is often the case that a speech recognition system must be incorporated.

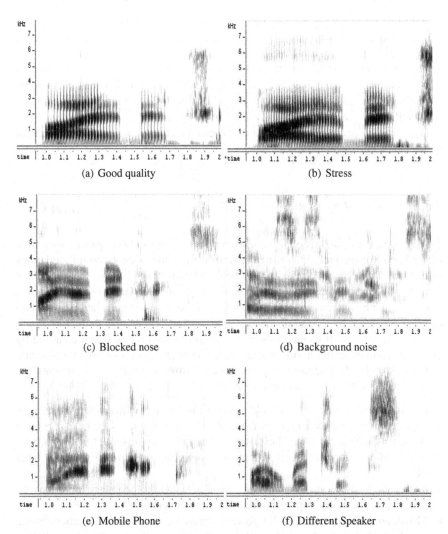

Fig. 3.5 Speech, examples of quality variation. The spectrogram show windowed (250Hz bandwidth) frequency versus time in seconds. Graphs (a)-(d) are from the same male with an Australian accent saying the word 'biometrics'. (a) **Good quality**: clear definition of vocal formant's. (b) **Stress**: higher frequency components and louder. (c) **Blocked nose**: some higher frequency components with less intensity. (d) **Background noise**: additional frequencies obscure the signal. (e) **Bad mobile phone call**: Mobile codecs compress the speech before transmission and so some frequencies are obscured. (f) **Different speaker**: a male with Canadian accent also saying 'biometrics'.

- **Mimics**: Some people are very talented at mimicking others voices. Whilst these individuals sound similar, the vocal signature still contains traces of the underlying physiology.
- **Relations**: People who are closely related to each other or are of the same gender and similar age may have very similar vocal physiology and speech style. Some testing results suggest that these individuals, whilst at an elevated risk of miss-identification, can still be distinguished from one another.
- **Age**: Speech changes with age for all people, however it is particularly apparent for males during puberty [18].

3.3.5 3D Facial geometry

Three dimensional face recognition uses various sensing technologies to determine the geometry of the face. This structure reflects the underlying skeletal foundations of the face more directly than can be obtained using two dimensional face data. Various sensing schemes have been used for acquisition, falling into three classes: passive sensing - stereo cameras that look for pixel to pixel correlation using two cameras separated by a fixed distance; active sensing - projecting a structured light onto the face (e.g. a grid) and noting the distortions in position that are caused by the facial geometry; and hybrid sensing that combines aspects of both passive and active sensing [3, 7]. The technology is still in its relatively early phase of adoption, with a small number of commercial vendors.

- **3D Rotation**: Depending on the geometry, reader information on range may be obtained from a single direction. This will cause occlusions as the head is rotated around its axis away from the camera. This effect is illustrated in Fig. 3.6 (f) where information on the left side of the nose is obscured and distorted by head tilt.
- **Noise**: Depending on the technology used to sense the geometry there may be spikes, pits and holes in the acquired surface geometry.
- **Movement**: The sensing of 3D geometry may be slower than a camera frame rate, hence it may lead to artifacts if the subject moves during acquisition. In Fig. 3.6 (c) the effect of the head moving during capture is shown.
- **Expression**: Facial expressions can radically change the geometry of the cheeks, mouth and nose. The effects of this are similar to two dimensional face recognition (see Sect. 3.3.2), however in some cases the effects are more drastic since the information available is only structural, not tonal. Figure 3.6 (e) shows the large effect that expression can have on distorting areas of the face and creating geometry holes.
- **Glasses**: Glasses cause the eye region to be occluded, since many range sensors are not able to sense through glass. The effect of this has been simulated in Fig. 3.6 (b).
- **Beards and hair**: 3D geometry systems are more capable of effectively using the structure of the jaw for recognition than two dimensional face recognition. As a

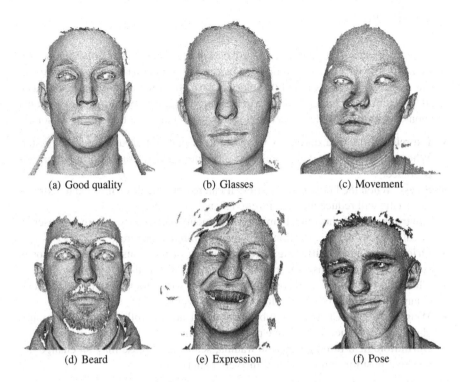

(a) Good quality   (b) Glasses   (c) Movement

(d) Beard   (e) Expression   (f) Pose

Fig. 3.6 3D Facial geometry, examples of quality variation: (a) **Good quality**: Even distance, neutral expression (b) **Glasses** (simulated): Effect of glasses on image. (c) **Movement**: Because of slow scan rates facial movement can skew source data. (d) **Beard**: Hair and beards can cause artifacts. (e) **Expression**: Mouth smiling and eyes closed. (f) **Pose**: Head rotated slightly to the left. Images (a)-(b) captured using Vivid 910 by Minolta, (c)-(e) FRGC v2 [19]. ©2008 IEEE

result, beards may affect performance. Figure 3.6 (d) shows the effect of facial hair on the image sensor - since the surface of a beard is uneven it can create holes in the data model.

- **Antiquity**: During growth years, the facial bones and structure change significantly. Furthermore, the muscles of the face become less tight, which leads to sagging. Both of these effects will alter the apparent geometry of the face.
- **Weight change**: Significant weight change can alter the 3D geometry of the face, although the geometry of facial features around the nose and eyes are less affected.

3.3.6 Vascular

The vascular network found just under the skin has been shown to be distinctive. Systems using veins for recognition are increasing in popularity. They work using a near infrared light transmitted or reflected through a biometric sample, such as a hand [4], palm [6] or finger [10], to map the pattern made by veins. As these systems are non-contact they are less susceptible to damage than most fingerprint sensors.

- **Exercise**: After and during exercise blood is pumped around the body faster. When this is the case, veins are more prominent and warmer, altering their appearance.
- **Stress**: When the body is under stress it can restrict the flow of blood to extremities. This will reduce the near infrared signature of the veins.
- **Environment**: A hot and humid environment, particularly where the user is sweating, may cause distortion of the near infrared signature.
- **Orientation and Positioning**: The positioning of the veins under the sensor is subject to three dimensional rotations which will distort their relative positions.
- **Clothing**: For palm vein recognition, the use of wrist straps or tight watches can change the amount of bloody flowing through veins.
- **Weight change**: Changes in subcutaneous fat after enrollment can potentially alter the appearance and relative position of veins.
- **Dermatological damage**: Recent trauma, scars and disease may all change the apparent position and location of the vein pattern,

3.3.7 Keystroke

Keystroke recognition is the use of inter-key timing and keystroke patterns for recognition [12]. It may also involve analysis of typing peculiarities or even word usage. Generally, a large sample of keystrokes is required in order to provide sufficient accuracy since individual keystrokes have high variability in timing.

- **Timing accuracy**: The timing of keystrokes needs to be at least at millisecond resolution. Therefore, recognizing keystrokes over a chat client is not practical due to inherent transmission delays (unless timing information is encoded into the character transmission).
- **Program Usage**: The type of data available for typing recognition depends on the way the computer is being used. For example, using email or a word processor may provide different keystroke patterns than when using a chat client.
- **Experience Level**: As a user gains experience with a keyboard or program their typing pattern is likely to change. In particular, keystroke timing becomes more precise and faster as familiarly increases.
- **Keyboard location and type**: The type of keyboard used (e.g. laptop, ergonomic or full-size) and where it is being used (e.g. desk, lap or train) will have an impact on performance.

3.3.8 Signature

Signature recognition is familiar to most people from its use with documents and credit cards. Dynamic, or online, signature recognition is a method that uses a device to capture the pressure and velocity of the pen movements in real-time for a more robust form of recognition [9].

- **Enrollment quality**: Signatures vary greatly in terms of their length and complexity, and hence the information content available for recognition varies. For example, someone with a short signature may be very easy to spoof.
- **Consistency**: Depending on the context when the signature was written, for instance when the writer is in a hurry, there can be a wide variation in the appearance, speed and pressure applied.
- **Surface angle**: The angle of the pen to the surface, depending on whether a person is seated or if they are standing, will affect the signature properties.

3.3.9 Hand Geometry

Hand geometry systems take an image of the hand inside a controlled enclosure and measure certain aspects of the geometry [16]. Distance measures include the lengths, widths and heights of the fingers, and the distance between knuckles.

- **Medical**: Arthritis can alter the appearance of the hands and joints, as well as make these devices difficult or painful to use.
- **Hand Placement**: The user is commonly required to align their hand using a series of pins. Misalignment of the hand can cause enrollment or acquisition errors.
- **Rings**: If rings are worn they may cause the finger geometry to be altered.
- **Weight change**: Weight gain or loss can affect the width of fingers and other hand dimensions.

3.4 Conclusion

This chapter has looked at biometric data from a number of different perspectives: the relationship between biometric entities, standards for biometric data storage, and quality issues affecting selected biometrics. The storage, management, protection and exchange of biometric data are important issues for consideration in any biometric implementation. The secure storage and proper management of biometric data has benefits beyond preventing external attack, as it can also help to enhance privacy protection and ensure smooth system operation.

Biometric standards have been under development for a number of years; however, until recently the adoption has been relatively slow. As the industry matures, the importance of standards compliance is being more widely appreciated. A particularly important development is the creation of a glossary of standard definitions. This will help ensure that customers, vendors and researchers are communicating using common terms and with accepted meanings. Efforts have been made to ensure that the terms used in this book (see Chap. 6) are consistent with the developing consensus around these definitions.

The old adage of 'garbage in, garbage out' applies particularly well to the relationship between data quality and matching performance. Therefore, measuring and monitoring the quality of biometric samples is of vital concern for the implementation and maintenance of any biometric system. The relationship of biometric quality scores and the suitability of a sample for authentication is an currently an active area of research. There are likely to be many further developments in techniques for quality control and assessment that will result in improved and more robust matching algorithms.

References

[1] Communications security establishment certification body canadian common criteria evaluation and certification scheme. http://www.cse-cst.gc.ca/documents/services/ccs/ccs_biometrics121.pdf (2001)

[2] Bimbot, F., Bonastre, J., Fredouille, C., Gravier, G., Magrin-Chagnolleau, I., Meignier, S., Merlin, T., Ortega-Garcia, J., Petrovska-Delacretaz, D., Reynolds, D.: A Tutorial on Text-Independent Speaker Verification. EURASIP Journal on Applied Signal Processing **2004**(4), 430–451 (2004)

[3] Bowyer, K., Chang, K., Flynn, P.: A survey of approaches and challenges in 3D and multi-modal 3D+ 2D face recognition. Computer Vision and Image Understanding **101**(1), 1–15 (2006)

[4] Cross, J., Smith, C.: Thermographic imaging of the subcutaneous vascular network of theback of the hand for biometric identification. Security Technology, 1995. Proceedings. Institute of Electrical and Electronics Engineers 29th Annual 1995 International Carnahan Conference on pp. 20–35 (1995)

[5] Ekman, P., Friesen, W.: Facial action coding system: A technique for the measurement of facial movement. In: Consulting Psychologists Press. Palo Alto (1978)

[6] Fan, K., Lin, C.: The use of thermal images of palm-dorsa vein-patterns for biometric verification. Pattern Recognition, 2004. ICPR 2004. Proceedings of the 17th International Conference on **4** (2004)

[7] Heseltine, D.: Face recognition: Two-dimensional and three-dimensional technique. http://www-users.cs.york.ac.uk/~nep/research/3Dface/tomh/PhD-Heseltine.pdf (2003)

[8] Hill, M.: Anat2310: Eye development. `http://anatomy.med.unsw.edu.au/cbl/teach/anat2310/Lecture06Senses(print).pdf` (2003)

[9] Jain, A., Griess, F., Connell, S.: On-line signature verification. Pattern Recognition **35**(12), 2963–2972 (2002)

[10] Lian, Z., Rui, Z., Chengbo, Y.: Study on the Identity Authentication System on Finger Vein. Bioinformatics and Biomedical Engineering, 2008. ICBBE 2008. The 2nd International Conference on pp. 1905–1907 (2008)

[11] Maltoni, D., Maio, D., Jain, A., Prabhakar, S.: Handbook of Fingerprint Recognition. Springer (2003)

[12] Monrose, F., Rubin, A.: Authentication via keystroke dynamics. Proceedings of the 4th ACM conference on Computer and communications security pp. 48–56 (1997)

[13] NIST: Published american national standards developed by incits m1 - biometrics. `http://www.itl.nist.gov/div893/biometrics/documents/April%206_%20FP_Published_INCITS_M1_Standards.pdf` (2007)

[14] O'Toole, A.J., Phillips, P.J., Jiang, F., Ayyad, J., Penard, N., Abdi, H.: Face recognition algorithms surpass humans matching faces over changes in illumination. In: IEEE Transactions on Pattern and Machine Intelligence, vol. 29, pp. 1642–1646 (2007)

[15] Phillips, P.J., Wechsler, H., Huang, J., Rauss, P.: The FERET database and evaluation procedure for face recognition algorithms (1998)

[16] Sanchez-Reillo, R., Sanchez-Avila, C., Gonzalez-Marcos, A.: Biometric Identification through Hand Geometry Measurements (2000)

[17] Scherer, K.R.: Effect of stress on fundamental frequency of the voice. In: The Journal of the Acoustical Society of America, vol. 61, pp. S25–S26 (1977)

[18] Sungbok, L., Alexandros, P., Shrikanth, N.: Analysis of children's speech. pitch and formant frequency. In: The Journal of the Acoustical Society of America, vol. 101, p. 3194 (1997)

[19] T.C. Faltemier, K.B., Flynn, P.: A region ensemble for 3-d face recognition. In: IEEE Transactions on Information Forensics and Security, vol. 3, pp. 62–73 (2008)

[20] Wildes, R.: Iris recognition. Biometric Systems: Technology, Design and Performance Evaluation, JL Wayman, AK Jain, D. Maltoni, and D. Maio, Eds. London: Springer-Verlag pp. 63–95 (2005)

[21] Yu, K., Mason, J., Oglesby, J.: Speaker recognition using hidden Markov models, dynamic timewarping and vector quantisation. Vision, Image and Signal Processing, IEE Proceedings- **142**(5), 313–318 (1995)

[22] Zhao, W., Chellappa, R., Phillips, P.J., Rosenfeld, A.: Face Recognition: A Literature Survey. ACM Computing Surveys **35**(4), 399–458 (2003)

Chapter 4
Multimodal Systems

Multimodal biometrics refers to the use of more than one source of information for biometric recognition [8, 9]. For example, a multimodal biometric system may use both iris recognition and fingerprint recognition to confirm the identity of a user. The use of multiple information sources helps to address some of the problems faced by real-world unimodal (also known as monomodal) systems, and multimodal biometrics will likely become increasingly common in future biometric deployments.

This chapter gives a brief, high-level introduction to the field of multimodal biometrics. The goals are to:

- Motivate the use of multimodal systems by outlining the primary advantages they offer over traditional systems (Sect. 4.1).
- Present the different approaches and modes of operation for multimodal systems (Sect. 4.2).
- Discuss information fusion and score combination (Sect. 4.3).
- Outline methods for the evaluation of multimodal systems (Sect. 4.4).

4.1 Advantages of Multimodal

The optimal biometric recognition system would be one having the properties of distinctiveness, universality, permanence, acceptability, collectability, and resistance to circumvention [9]. No existing biometric system simultaneously meets all of these requirements, however the use of more than one biometric can help lead to a system that is closer to these ideals. The advantages of multimodal systems stem from the fact that there are multiple sources of information. The most prominent implications of this are increased accuracy, fewer enrollment problems, and enhanced security.

4.1.1 Accuracy

The most immediate advantage of multimodal authentication is increased recognition accuracy. Multimodal systems fuse information for more than one source, each of which offers additional evidence about the authenticity of an identity claim. Therefore, one can have more confidence in the result. For example, consider two people who coincidentally have a similar facial appearance. In this case, there is a potential risk of a false accept for a system based purely on face recognition as a relatively high match score may be achieved when matching one against the other. However, if the same system also included fingerprint matching, it would be very unlikely that any given two people would have similar faces *and* similar fingerprint patterns. Therefore, the ability of the system to distinguish between people is increased significantly.

The previous example illustrated the use of multimodal matching for a verification system (one-to-one match). However, multimodal biometrics is particularly useful in an identification situation, as this involves many matches. Assume a fingerprint matching algorithm has a false match rate of 0.01%, which would be considered a high-accuracy system for fingerprints. With a database of 1,000,000 fingerprint images, one would expect approximately $0.0001 \times 1,000,000 = 100$ false matches for every identification query. In most cases this would be considered unacceptable performance, as it would be very laborious to manually examine all 100 false matches in the hopes of finding a correct one. Assume that an iris image is also stored for every person enrolled, and an iris matching algorithm is available that also has a false match rate of 0.01%. Under some reasonable assumptions, the probability of a fingerprint falsely matching an enrollment is independent of the probability that an iris will falsely match the same person, so the two probabilities can be treated as statistically independent. Therefore, the individual error rates can be multiplied together to find the expected error rate for the combined system.[1] The expected number of false matches for a multimodal identification query is $0.0001 \times 0.0001 \times 1,000,000 = 0.01$. This means that only one in every 100 queries would return a false match, which would be considered acceptable for most practical applications. This is an improvement of several orders of magnitude over using a fingerprint alone, and demonstrates the power of multimodal combination to boost the performance of identification systems.

4.1.2 Enrollment

Another advantage of multimodal systems is that they address the problem of non-universality, where a portion of a population has a biometric characteristic that is missing or not suitable for recognition. For many multimodal systems matching can be conducted even when one of the samples is unavailable or excessively poor

---

[1] This makes the assumption of an 'and' combination policy, rather than an 'or' policy.

quality. This will reduce the failure to enroll rate significantly. For example, consider the case of a person with damaged vocal cords who cannot speak, but would like to enroll in a system that uses voice authentication. In this case, their ability to enroll using an alternative biometric (such as a fingerprint) is a necessity.

Poor quality data is a common cause of enrollment errors, and ultimately false accepts and false rejects. The multimodal approach provides the system with a "second chance" to obtain or match a sample of sufficient quality, thereby increasing the robustness of the system.

4.1.3 Security

Multimodal systems have increased resistance to certain types of vulnerabilities, in particular spoof attacks. A spoof attack is where a person pretends to be another person by using falsified information. In the context of biometric systems, this involves creating and presenting an artificial representation of another person's biometric. For example, Japanese researchers have demonstrated how to create fake fingerprints using a commonly available material (gelatin) that has some success at fooling commercial fingerprint recognition systems [6]. The advantage of multimodal systems is that an attacker would have to be able to spoof two different biometric modalities simultaneously, which would be significantly more challenging.

4.2 Types of Multimodal Systems

There are several different ways that multimodal systems can be constructed, based on the sources of the biometric information and the way the system is designed. The term 'multimodal' sometimes refers specifically to the case where two or more different biometric modalities are in use (such as face and fingerprint), while the term *multi-biometrics* is more generic. Multi-biometric systems includes multimodal systems, as well as a number of different configurations.

Multimodal systems can be characterized by their sources of information or their organization.

4.2.1 Sources of Information

At a fundamental level, all multi-biometric systems collect and combine information from a variety of sources. However, the sources of the information differ from system to system. The following are the most common approaches:

Multiple modalities   This refers to the situation where different biometric modalities are used, such as faces and fingerprints. The primary advantage of this approach is that it maximizes the independence between the samples. Therefore, a problem authenticating with one biometric is unlikely to impact authentication with the other.

Multiple characteristics   This is the use of different instances (characteristics) of the same biometric. For example, one can match both thumbs for a fingerprint system, or the left and right eye for a retina system. In most cases, different biometric characteristics will have a high degree of independence. Furthermore, implementation is usually simple and cost effective because the same sensing equipment and matching algorithms can be used for each instance.

One characteristic, multiple sensors   This uses multiple captures of the same biometric from different sensor types. For example, it may use both 2D and 3D face data. Another example is multi-spectral approaches, such as capturing a biometric with both the visible spectrum and infrared spectrum. The disadvantage of this approach is that if a biometric is not suitable for recognition (e.g. missing or damaged), the performance benefits of multiple captures will be minimal.

One sample, multiple algorithms   This is the combination of multiple algorithms used to match the same sample. For example, different vendor face matching engines are applied to the same images. The advantage of this approach is where the algorithms have been developed independently, each will have its own strengths and weaknesses, and therefore may contain complementary information. However, since both algorithms are applied to the same data, there will be a degree of correlation between the results, and both algorithms will struggle with poor quality input.

Multiple impressions   This technique uses multiple impressions of the same biometric characteristic. For example, multiple faces from a continuous video stream can be extracted and matched using a single matching engine. Another example is the integration of signals from multiple samples acquired at discontinuous intervals over an extended period of time.

Soft biometrics   Multimodal systems can also use information from "soft" biometrics traits like height, weight, and eye color. These traits may have considerable variation in both the quality of the acquired data and temporal variation, so are of little value when used in isolation. However, some research has shown accuracy improvements when used in combination with other biometric traits [1, 3].

4.2.2 Modes of Operation

From the point of view of system flow, there are two common ways for a multimodal system to operate: in parallel, or in serial.

In serial mode, the various acquisitions and matching stages are conducted one after another. An example of this is a cascaded system, where if the user fails one biometric system, they use another biometric system, and the final output score is the fused scores of both. For example, the user first uses a fingerprint sensor, if this fails face recognition is used, and if this fails a hand print is required. The advantage of such a system is that many users will only need to use one sensor, streamlining the verification process. Some speaker verification systems use a variant of this where challenge response questions are asked until the user reaches a specified authorization level or the call is terminated.

For parallel systems everything is done somewhat simultaneously. For example, an ATM may require a person to have their fingerprint captured while they are looking at a camera for face recognition. There are security advantages to systems designed in this way because they are difficult to spoof. However, they may be more inconvenient and cumbersome to use from the user's point of view.

Combination Policy

There is a wide variety of ways that a multi-biometric system can be put together; either through logical combinations of match conditions or the application of various algorithms to combine scores. There is not one best combination technique for all circumstances, and deductions about what seems best from common sense may lead to poor outcomes in operation. The best advice is to choose a selection of potential combination techniques and run a series of scenario trials to determine which one will really perform the best for the desired outcomes (see Chap. 5).

4.3 Combination Techniques

Techniques for biometric fusion fall into three general categories, depending on the stage at which the combination is conducted: *feature fusion* combines low-level distinguishing features, *score fusion* makes use of multiple match scores, and *decision level fusion* logically combines accept/reject matching decisions.

4.3.1 Feature Level Fusion

As a general principle, better performance can be expected from feature level fusion than from score and decision level fusion. The reason for this is that the most information is available, which may be lost when fusion is conducted at higher levels. However, there are a few difficulties that make feature level fusion challenging and less common than other methods. First of all, the feature sets for the different information sources may be entirely different. For example, combining fingerprint and face features into a single model would not be a straightforward task as the matching algorithms (or distance metrics) used to compare these features are entirely different. Secondly, by simply concatenating features from different sources one would suffer from the problems associated with high-dimensional features spaces (known as the "curse of dimensionality"). Finally, for most commercial systems the feature data will be proprietary and not accessible to the system integrator.

One example of feature level fusion is the combination of fingerprint minutiae data from two different minutiae extraction algorithms. Both algorithms will have their own strengths and weaknesses, but when combined they will result in a more robust model of the fingerprint's minutiae. When this combined data is used for matching, better results are expected than when combining the final scores generated by the two algorithms because these are based on weak templates.

4.3.2 Score Level Fusion

Multimodal fusion is an active area of research, and the main emphasis tends to be on fusion at the match score level. There are several advantages to this approach:

- In general, score level fusion achieves better results than decision level fusion. The reason for this is that useful discriminative information (e.g. the confidence of a decision) is lost when the match score is disregarded.
- Match score fusion can be applied to most of the multi-biometric schemes discussed in Sect. 4.2. For example, it can be used to combine the scores of two fingerprint pairs, or the scores of a face match and a fingerprint match. Feature level fusion (discussed above) can be more difficult to apply in some circumstances.
- There is no need to have knowledge of the underlying algorithms or have access to feature information, as is the case for feature-level fusion. This is an advantage because it makes implementation easier. Furthermore, many commercial systems have proprietary formats for the storage of the feature data, so the information cannot be accessed directly.

There are two approaches to combination at the similarity score level: classification and score combination.

4.3.2.1 Classification

The idea of the classification approach is to consider verification as a classification problem with two classes: 'Accept' and 'Reject' [2]. Each authentication is represented by a feature vector that is composed of similarity scores from the various sub-systems. For example, assume face, fingerprint, and voice algorithms output similarity scores of 89, 76, and 58 respectively. This would lead to the feature vector [89, 76, 58], which would have an associated label of 'Accept' or 'Reject'. Training samples are collected, and machine learning or pattern recognition techniques (such as neural networks, support vector machines, decision trees, etc.) are used to build a model to distinguish the two classes. This classification model is applied to unseen data to make a verification decision.

4.3.2.2 Score Combination

Score combination involves taking several scores and applying a formula to combine them into a single score. Some examples include adding the scores together, taking the average, or selecting the minimum or maximum score. Several studies have concluded that the sum rule (a weighted average) is the best option due to its simplicity and high performance [4, 7].

One of the main issues that must be addressed for score level fusion is known as *normalization*.[2] The issue is that the similarity scores from different algorithms may not share the same underlying properties or score range. Therefore, simple approaches such as taking the average of two or more scores usually cannot be applied without first performing normalization. This can be illustrated with the following example. Consider Algorithm 1 that generates scores in the range [0...1] and Algorithm 2 that outputs scores in the range [0...100]. Obviously, taking a direct average will not work as the Algorithm 2 score will dominate the result. However, even when the score range is the same, the distribution of scores in this range (e.g. the mean and variance) may be entirely different. Score normalization is the process of transforming match scores from different sources into a standard distribution (both range and shape). Once this has been done, the individual scores can be compared against each other and combined accordingly.

When combining scores that have been generated from different biometric characteristics, the scores will have a high degree of independence. For example, if a person achieves a high fingerprint match score, this usually will not imply how well their face image will match. However, in some cases the match scores from different systems will be related to each other in some way. This will be particularly true when multiple algorithms are being applied to the same input data. This is illustrated in Fig. 4.1, which shows the results of applying two fingerprint matching algorithms to the same data. The effect of this is that the separation between the genuine and impostor scores is increased. A linear discriminant function is included that is able to distinguish between the two classes better than one based on the scores from only one algorithm (i.e. a horizontal or vertical line).

---

[2] Note that this is different from the concept of normalization for identification systems, which attempts to maximize the distance between vectors in the feature space (see Sect. 7.2.1.2).

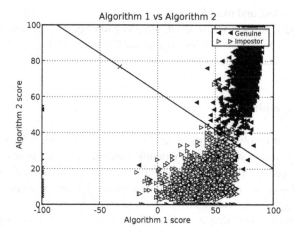

Fig. 4.1 Multi-algorithmic analysis of fingerprint matching results using a FVC 2002 data-set [5]. The plot is of a minutiae algorithm against a non-minutiae algorithm. The straight line through the points shows a separation between impostors and genuine points that is better than can be achieved using a single algorithm. The decision function for the line is that a match is genuine if $65*(Score_{Algorithm2}) + 17*(Score_{Algorithm1}) > 3752$, leading to a FNMR of 1.6% and a FMR of 0.2%.

4.3.3 Decision Level Fusion

Decision level fusion is the highest level combination possible in the sense that all information about the matching process has been extracted except for a binary decision. In some circumstances, the raw match score is not available, such as for commercial systems that output "accept" or "reject". In this case, no information is known about the confidence of the decision. In other words, there is no way to distinguish between a strong match and a borderline match.

There is limited analysis that can be conducted for decision level fusion, and consequently the combination schemes are less sophisticated than for other modes of fusion. The most common approach is to use a majority voting scheme. For example, if there are three matching engines, the final decision is that which at least two engines agree on. When there is an even number of matching systems a voting scheme can lead to inconclusive results. More complex rules can be constructed using heuristics, such as "accept the user if any 6 out of the 10 fingerprint pairs match".

4.4 Evaluation

For feature and score level fusion, the evaluation of multimodal results is generally the same as the evaluation of other biometric systems (see Chaps. 7-9). The reason for this is that the output of multimodal systems is essentially the same as that from a unimodal system: a similarity score for verification systems (see Sect. 7.1), or a rank for identification systems (see Sect. 7.2). Therefore, the same modes of analysis can be conducted. In particular, false accept and false reject rates are used to report verification performance (e.g. ROC curves), and identification rates are used to quantify identification system performance (e.g. CMC curves). For decision level fusion match scores are not available, so performance rates are reported at a fixed operating point.

One of the primary advantages of multimodal systems is their ability to reduce enrollment errors. Therefore, an emphasis on failure to enroll rates should be an integral part of any multimodal evaluation.

4.5 Conclusion

This chapter has provided a high-level introduction to multimodal biometric solutions. The use of multi-biometric systems for enhancing system performance at both an algorithmic and system level is becoming increasingly important due to its numerous benefits over unimodal systems.

There are some disadvantages of multimodal systems, as they may be more expensive and complicated due to the requirement of additional hardware and matching algorithms, and there is a greater demand for computational power and storage. From a user's point of view, the systems may be more difficult to use, leading to longer enrollment and verification times. Furthermore, there are interoperability challenges related to the integration of products from different vendors. Despite these challenges, the field continues to be an active area of research because of the potential benefits of increased accuracy and security, and fewer enrollment failures.

As new biometric techniques are introduced they can often be combined with existing biometrics. Some recent examples include: iris and face, skin-texture and face, and skin pores and fingerprints. However, regardless of how systems are fused together, the final result is still a matching module that outputs a single score on which a decisions are made. Hence, the techniques described in Part II of the book apply equally to multi-biometrics systems as they do to unimodal systems.

References

[1] Ailisto, H., Vildjiounaite, E., Lindholm, M., Mäkelä, S.M., Peltola, J.: Soft biometrics-combining body weight and fat measurements with fingerprint biometrics. Pattern Recognition Letters **27**(5), 325–334 (2006)
[2] Ben-Yacoub, S., Abdeljaoued, Y., Mayoraz, E.: Fusion of face and speech data for person identity verification. IEEE Trans. on Neural Networks **10**(5), 1065–1074 (1999)

[3] Jain, A.K., Dass, S.C., Nandakumar, K.: Soft biometric traits for personal recognition systems. In: Proc. of ICBA, pp. 731–738 (2004)

[4] Kittler, J., Hatef, M., Duin, R.P.W., Matas, J.: On combining classifiers. IEEE Trans. Pattern Anal. Mach. Intell. **20**(3), 226–239 (1998). DOI http://dx.doi.org/10.1109/34.667881

[5] Maio, D., Maltoni, D., Cappelli, R., Wayman, J.L., Jain, A.K.: FVC2000: Fingerprint verification competition. IEEE Trans. Pattern Anal. Mach. Intell. **24**(3), 402–412 (2002)

[6] Matsumoto, T., Matsumoto, H., Yamada, K., Hoshino, S.: Impact of artificial gummy fingers on fingerprint systems. In: Proc. of the SPIE, Optical Security and Counterfeit Deterrence Techniques IV, vol. 4677 (2002)

[7] Ross, A., Jain, A.: Information fusion in biometrics. Pattern Recognition Letters **24**(13), 2115–2125 (2003)

[8] Ross, A., Jain, A.: Multimodal biometrics: an overview. In: Proc. of the 12th European Signal Processing Conference, pp. 1221–1224 (2004)

[9] Ross, A.A., Nandakumar, K., Jain, A.K.: Handbook of Multibiometrics. Springer (2006)

Chapter 5
Performance Testing and Reporting

The evaluation of system performance and assessment of overall system reliability are important to the implementation and use of all biometric systems. Almost all biometric systems are 'black boxes' to their users, and it is only through appropriate testing that a system's accuracy and performance can be determined. However, this testing process is not always straightforward, as there are many factors that influence performance, and published error rates often provide only approximate guidance. Issues related to the misunderstanding of test results are frequently the reason why some biometric implementations have failed to meet expectations. Properly conducted testing can provide crucial information for the planning of a system's functionality, and the introduction of controls to maximize results.

The goals of this chapter are to:

- Discuss the evaluation process (Sect. 5.1).
- Describe the components of a test plan (Sect. 5.2).
- Introduce the different types of evaluation (Sect. 5.2.1).
- Present the fundamental techniques for assessment, including establishing ground truth and data size (Sect. 5.3.1).
- Discuss issues related to the test set (Sect. 5.3).
- Present the elements of proper reporting of test results (Sect. 5.4).

5.1 Introduction

An evaluation can choose to assess a variety of technical measures including acceptability, availability, maintainability, vulnerability, privacy, security, cost/benefit trade-offs, standards compliance, reliability, usability and accuracy. This chapter is primarily concerned with accuracy and security. Fundamentally, this is providing methods to assess a biometric system's capability to identify a correct user and reject an incorrect user. These methods must ensure that the calculated results from the trial will provide an accurate reflection of performance when the system is in operation.

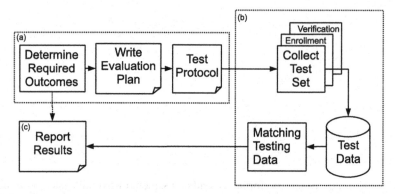

Fig. 5.1 Overview of the testing process. a) Determining the required outcomes and creating an evaluation plan. b) Collecting the test set and conducting the matching - this differs depending on the evaluation type. c) Reporting the final results.

It is vital that the outcomes are well defined prior to conducting the tests in order to help ensure that the final report answers the correct questions. The evaluation plan needs to set out how the testing process will meet these objectives and may include: resources, scheduling, environmental controls, the data acquisition process and test subject training. The next phase is to collect the biometric data and undertake biometric matching. Furthermore, the results must be reported in a way that is accurate and concise, but also understandable to its intended audience. The testing process is illustrated in Fig. 5.1.

Effective and accurate performance evaluation can be complicated and costly. A primary reason is that a useful *test set*, which is collection of biometric data, usually requires recruiting a large group of people, known as a *test crew*. The test crew should reflect the target population, so may need to include a wide spread of demographics. Whilst every person's biometrics are different to varying degrees, groups of people related by birth, ethnicity or profession will often share certain aspects of physiology. The effect of these shared characteristics varies depending on the biometric, but knowing in advance if any segment of the population will have problems with enrollment or verification is critical. For each person there will be at least one enrollment, and one or more verification events over an extended period. It is through a process of repeated transactions, from many different people under a wide variety of conditions, that statistical confidence is built in the genuine and impostor score distributions (see Chaps. 2 and 7). Ideally, the testing process should illuminate the "true" system accuracy for a given application, and determine the biases that exists across the entire range of relevant users and environments.

5.2 The Test Plan

In conducting any type of evaluation it is important to set out clearly at the start what is to be evaluated and why. Common cases include the comparison of several algorithms on a standard test set, support for a business case, the prediction of error rates as enrollment numbers increase, or setting and tuning system parameters.

The *test plan* needs to set out a path that involves identifying the the test target(s), determining what data needs to be collected, creating a schedule and resource plan, specifying how the data will be matched and describing the way results will be presented. Where multiple vendors are to be compared, testing protocols are generally required. These protocols help assure the participants that the testing process is fair and unbiased. The protocol specifies how the matching will take place, standards for interfacing with the engine(s) or extracting matching data, and what conditions are being put in place to ensure there are not opportunities for cheating. The test set needs to be de-identified so that it is not possible to deduce a genuine from an impostor match by simply looking at file names or other structures (such as having genuine and impostor matches in separate directories).

It is recommended for most evaluations to allow for *run-up tests*, which will help identify potential problems and ensure smooth operation. This involves providing a small test set that can be used by the vendors to tune their system and make sure they can handle the input format and type. The test set should be reflective of average quality that will be used for the actual test, but not included in the real test set.

5.2.1 Evaluation Types

Evaluations can be conducted with different degrees of rigor and control depending on the required outcomes. The testing procedures are defined in a series of ISO standards [3], which have been developed largely based on a best practice document released in 2002 [6]. Each type of testing is appropriate for a different stage in the biometric life-cycle (see Table 5.1).

Technology Evaluation		**Scenario Evaluation**		**Operational Evaluation**
Aim: To short list technologies, obtain accuracy estimates and generate comparative results.	→	Aim: To determine realistic accuracy levels and provide initial tuning for the system.	→	Aim: To tune a system and ensure that false accepts and rejects are as expected. Also determine system weaknesses and how they might be addressed.

Table 5.1 When to use the different evaluation options.

5.2.1.1 Technology Evaluation

A technology evaluation focuses on testing the core of the biometric system: the recognition algorithm. This is achieved by conducting matches in an offline or batch mode. A set of 'standard' users are enrolled, and a number of impostor and genuine matches are undertaken. Technology evaluations are often undertaken to determine the state-of-the-art in a field as part of a public vendor comparison, or as part of short-listing the best technology for a particular project. Since the test set is consistent for all algorithms under evaluation, the performance results are repeatable and directly comparable. Well known examples of technology evaluations include the Face Recognition Vendor Tests (FRVT) [9] and the Fingerprint Verification Tests [5]. In some cases, the testing data is specifically acquired for the evaluation, and in other cases publicly available benchmark databases are used. Results from technology evaluations can be used for estimations of expected accuracy in production only when the test size is sufficiently large, and the data used is representative of the data that would be seen operationally.

5.2.1.2 Scenario Evaluation

Scenario evaluations undertake the testing in a simulated, but controlled, environment that is as close to operational as practical, given cost and time considerations. This means that the full end-to-end matching system, including all the relevant components needed to conduct the match, are ideally used for both enrollment and verification. In this way, the real use of the system is emulated, including user behavior and environmental factors. The evaluation should also involve the testing of boundary cases of sample quality, such as those listed in Sect. 3.3. In some cases the data for a scenario evaluation may already exist, such as for face recognition in applications like passport agencies and driver's license authorities. One of the best publicly available examples is tests undertaken by Mansfield et al. in 2001 on face, iris, fingerprint, face, vein and voice [7].

Scenario Testing During Algorithm Development

Technology evaluations are often used as the primary way to assess progress in improving accuracy during the development of biometric algorithms. However, a focus on technology testing can lead to poor performance in real world applications as there will inevitably be factors present in operation that were not modeled in the test set. A well run scenario evaluation can discover these factors, thereby giving a better estimate of real-world performance. It is suggested that, where possible, algorithm developers focus on scenario evaluation as the primary way to assess accuracy. This can be accomplished using a library of different testing scenarios. The performance of leading biometric solutions has benefited immeasurably from tuning their systems using realistic scenario testing.

5.2.1.3 Operational Evaluation

Once a system is in operation it is necessary to ensure that it is operating as expected, and is appropriately tuned. Real operational systems typically have problems that are not anticipated during technology or scenario testing. The tuning of operational systems is an important part of final system commissioning when moving from a pilot to production. Once in production operational testing may be undertaken periodically or continuously. Few operational tests are currently released publicly, since in most cases this data is considered sensitive and confidential. However, it is possible that in the near future audit requirements will necessitate greater transparency on publicly facing system, particularly those used by governments.

Unexpected Operational Results

When capacitance based fingerprint sensors were first introduced they had problems with the high electromagnetic fields generated from fluorescent lighting. The induced electric fields in users prevented or distorted fingerprints during the capture process. Apparently, the original testing had taken place under non-fluorescent lighting, and so had not demonstrated this problem.

5.2.2 Elements of an Evaluation

To ensure testing results are fair and accurate there is a wide variety of criteria that should be considered in the test plan. Some criteria are sufficiently important that they should be addressed in every test plan (primary test elements), whilst other criteria depend on the type of evaluation and intended application (secondary test elements).

5.2.2.1 Primary Test Elements

The biometric samples used for the test should approximate, in every way practical, the type and quality of data that the system will use in operation. The demographics and user behavior should be in similar proportion to those in the target user population, the data must be sampled from the same sensor types, and data should be captured from similar environments, all with the same average quality. Where the biometric system will be expected to match over a long period of time, equivalently old enrollments also need to be included in the test set.[1]

The observed accuracy results are highly dependent on the number of samples, and so can be highly biased and misleading if not properly reported. Figure 5.2 shows an example where, for small test sizes, an inferior algorithm appears to be better than a rival. One way to measure the certainty of a result is known as a confidence interval, and a goal of testing is to ensure that enough test data is used to make this interval is as small as possible. This is discussed below in Sect. 5.3.1.

When designing tests, it should be kept in mind that the more matching data available, the more complete the analysis that can be undertaken. Ideally all data is matched against all other data. This allows an examination of group-level effects (inter-class and intra-class differences) as well as allowing for an assessment of both verification and identification accuracy.

The type and frequency of enrollment or acquisition errors need to be recorded and reported. If errors are excluded from reporting this can significantly bias results. For example, an algorithm could be selective and only match those samples that it considers to be of sufficient quality. This could lead to high apparent accuracy whilst, in reality, rejecting many genuine attempts.

Errors in the ground truth (i.e. mislabeled data) will obviously lead to incorrect statistics. In large data sets, it is commonly the case that some data is mislabeled. Tests to look at outlier matches (very high impostors or very low genuine) will often catch many of these cases (see Sect. 5.3.3).

To ensure the failure modes of a system are understood, 'extreme' test cases should also be used. For instance, this may include matching identical twins or closely related individuals of the same gender and age. Extreme environmental conditions, such as lighting or excessive noise, might also be relevant depending on

---

[1] It is often the case for new biometric types that it is difficult to say how stable they will be over time as there may be limited longitudinal data on which to build a test population.

the target application. Examples of reasons for poor quality samples for the main biometrics can be found in Chap. 3.

Biometric data is inherently sensitive and has privacy implications if misused, so it should be protected by being de-identified during any testing process, and where practical stored securely using encryption. Permission forms should be obtained and achieved from test subject indicating informed consent for the intended use.

5.2.2.2 Secondary Test Elements

There are a variety of other criteria that may be relevant for testing, including:

- **Vulnerability:** It is frequently important to understand how a deliberate impostor might fool the system. The results of a vulnerability study should go into a risk management plan ensuring that appropriate mitigations are considered. Vulnerabilities are the subject of Chap. 12.
- **Fraud:** In identification systems, it is important to undertake a large scale match in such a way as to identify cases of multiple fraud. In many databases these cases may not be known in advance, and require a skilled operator to examine the top matches to see which are actually the same individuals. Methods for detecting fraud, and estimating levels of fraud, can be found in Chap. 10.
- **Open-set:** Some tests should be conducted where the individuals being tested are not in the enrollment gallery. This ensures that the algorithm is not making implicit assumptions about a presented individual being enrolled in the system, as such assumptions can artificially enhance performance.
- **Blindness:** Where the evaluation data has been used previously, or is known to the algorithm designers, it compromises the integrity of the testing process as the matching algorithm could cheat by being specially tuned to perform well on a particular data set. In other words the tested systems should see the test data for the first time during the evaluation.
- **Validation**: There are many subtle errors that can be be unknowingly introduced in the techniques used for the creation of performance figures and graphs. The software used to output the results should have been validated, either through its use in many other tests, or by having been checked for correctness against a sample set where the results are known.
- **Metadata:** To determine what biases exist in the matching process, as much metadata about the environment, individuals and the templates should be collected as possible. This allows data mining techniques (see Chap. 9) to be used to establish performance trends and biases. For example, a test might be designed to determine if 18-25 year old men carrying UK passports are easier to distinguish than the rest of the population using a particular sensor and matching algorithm.
- **Publicity:** The results of testing are often commercially sensitive for those organizations taking part. Before the testing process begins, agreement must be reached about what information, if any, will be released to the vendors and what will be released publicly.

- **Repeatability:** Ideally, the results of tests should be reproducible and auditable. This can be achieved by archiving the biometric samples, raw match scores and metadata. This can also be of assistance when assessing the relative performance of future upgrades.
- **Scalability:** Tests are usually undertaken with a limited number of samples. When a larger number of people are using the real system the accuracy may change, particularly with respect to identification results. Query times should also be tested and recorded, as if they do not grow linearly with the databases size, there may be future scalability problems.

Synthetic Biometrics

For many biometrics it is possible to synthetically generate biometric samples, and this can allow scalability to be evaluated by the generation of very large test sets. For instance, synthetic fingerprint generation [2] is used as part of large scale fingerprint tests [5]. However, the use of such sets should be treated with caution since the data may not be reflective of the actual acquisition process, and hence can easily lead to a systematic bias. Where artificial sets are used, they must be validated against results obtained using real sample data.

- **Storage artifacts:** Where biometric data is stored or transmitted before being matched it may be the subject of lossy compression, such as for images stored as JPEG files. Compression introduces noise artifacts into the sample data proportional to the level of compression. Excessive compression may impact the accuracy of the matching algorithm.
- **Subject behavior:** The behavior of the test subjects can have a big impact on performance, particularly in surveillance systems. The impact of this on the test plan is two fold. Firstly, the test script should indicate how the testers interact with the system. For instance, it should address questions such as: are the people familiar with technology, are they in a hurry, will they be carrying other objects such a luggage, etc. Secondly, during testing the subjects should be observed and anomalous behavior recorded.
- **Reliability:** Biometric matching, particularly for identification on large data-sets, can require significant processing and memory resources. Software issues, such as memory leaks or crashes, may need to be tested to ensure that the software does not fail during operation.
- **Feature support:** Matching software might include other advanced features to assist with the matching process, including search optimization using binning, multiple watchlist management, group level thresholds, template adaptation and data comparison/enhancement tools.

5.2.3 Benchmarking Against Human Performance

Biometric systems are sometimes installed as replacements or supplements to human jobs such as the checking of identity documents. Where this occurs it is reasonable to ask how well the biometric system compares to a human at the task of recognition. For instance, whilst people are generally considered to be very good at recognizing familiar faces, unfamiliar recognition is much more difficult, and recent results indicate that several modern algorithms surpass human performance [8]. However, people can draw on aspects of suspicious behavior that are usually outside the scope of a biometric algorithm. Understanding the performance difference between human and machine recognition can be an important part of risk analysis and establishing a business case for the introduction of biometrics. It is also increasingly important in the area of forensic evidence (see Sect. 10.2).

When comparing automated algorithms against humans, one must consider each person as a different 'recognition algorithm'. Some people are inherently better at pattern recognition than others, and each person may exhibit different bias depending on personal experience. In general, people can be trained to boost their recognition performance. This has been used effectively in fingerprint recognition for many years, but remains an area of research for other biometrics.

5.3 The Test Set

The process of generating match scores to determine genuine and impostor distributions is the essential task of biometric testing, as these distributions form the basis of all performance statistics. There are two main requirements necessary to ensure the relevance of the test set: there must be enough biometric samples to make statistically valid inferences about the system's accuracy, and the data used to generate these statistics must be similar to the population and environment of the target system.

5.3.1 Test Size and Results Confidence

As a result of the fact that the testing process is merely sampling from the pool of potential matches, the results from evaluations are only as precise as the test set is large. The lack of precision is due to *sampling error,* and its effects can be shown by using *confidence intervals.* Confidence intervals show the precision of the reported test result (see Sect. 7.3.3). Without knowing the confidence level of the results, one algorithm might appear to be better than another, yet could in reality be poorer (see Fig. 5.2). This discrepancy can usually be uncovered by using larger test sets.

Determining how many genuine and impostor matches are needed for a statistically significant result, or alternatively determining when to stop testing, is still an area of active research [11]. However, it is known that the more accurate the system, the more data that will be required to determine its error level with a certain degree of confidence. The standard recommendation is to test with the largest population set that can be reasonably managed, as the larger the test the more accurate the results. Two commonly used heuristics for sample size estimation are the Rule of 3 [4] and the Rule of 30 [10], which are explained in more detail in Sect. 7.3.6.

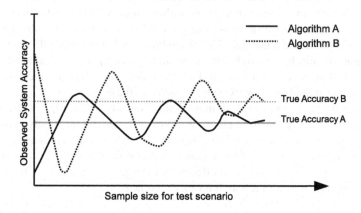

Fig. 5.2 Sample size versus observed system accuracy for two systems. As the number of test samples increases, the observed accuracy converges towards the true system accuracy. Note that at some small sample sizes Algorithm A appear better than Algorithm B.

5.3.2 Ground Truth

When a match is conducted, it is either a genuine match (the samples are from the same person) or an impostor match (the samples are from different people). The knowledge of whether or not a match is genuine or impostor is referred to as *ground truth*. In other words, ground truth records the actual state of nature, regardless of the match result. Collecting ground truth during an evaluation is of the utmost importance as performance analysis cannot be conducted without it. If ground truth is not available, one cannot know if an error has occurred (e.g. a false accept) so performance accuracy cannot be established.

Ground truth is usually recorded by labeling samples, as opposed to labeling matches, because this information can be used to determine the ground truth of any match. The key information that must be recorded is a unique identifier of the person from whom the biometric was captured. In the case that multiple characteristics are obtained from the same person (e.g. all ten fingerprints), this information must also be recorded. A good strategy is to encode this information into the sample file names.[2] For example, '0129_L_E_01.jpg' may be an image of an iris using a convention where the first four characters identify the person, 'L' or 'R' specifies which eye was captured, 'E' or 'V' denotes an enrollment sample or verification sample respectively, and this is followed by an attempt number. In order to establish ground truth, all that needs to be done is to compare the person ID and the characteristic for the two samples being matched.

There are some situations where establishing ground truth is very difficult. For example, it is often difficult to determine ground truth for an operational evaluation, because all that is known is the "claim of identity", which does not necessarily correspond to the true identity. When this is the case, matches must often be confirmed manually, using additional information when available. Another challenging situation is surveillance applications, where a genuine event can occur, but not have any corresponding samples. This happens when a target walks through an area under surveillance but no images are captured. Knowing that this event has occurred is an essential part of ground truth as it is required in order to compute detection rates. Establishing ground truth for surveillance applications is discussed in detail in Sect. 11.3.2.

Existing Evaluation Data

For some evaluations the testing data is not collected, but already exists. For example, consider the case where a driver's license authority would like to determine how well a face matching algorithm would perform for their existing database (see Sect. 10.1). In this case, the natural test data is a subset of actual driver's license photos, and ground truth is established using the existing records. However, one cannot assume perfect data integrity, as some degree of labeling errors and fraud is inevitable. It is important to conduct an assessment of data integrity as the results of an evaluation are only as sound as the ground truth used. This is discussed further in the next section.

5.3.3 Errors in the Test Set

In any large test set there is the potential for mistakes in the labeling (or ground truth) of the data. The larger the test set, the more likely it is that some data is incorrect. Where the number of mistakes is much smaller than the error rate being measured, the presence of the labeling errors simply causes a small amount of noise in the results. However, where the number of errors is large it will cause a serious bias in the test results. This can be dealt with either before or after data collection.

---

[2] For evaluation trials, this information should not be included in file names as it would allow an algorithm to establish ground truth independent of the samples.

5.3.3.1 Before Data Collection

Setting up data collection to minimize the chance of errors is the best method to ensure clean ground truth. This means making sure that any data entered is sanity checked and confirmed before storage. A common cause of errors is where a test subject forgets their unique ID, or mistakenly enters the wrong one during enrollment or verification. The risk of this can be reduced by the use of identity tokens (like smart cards) or good interface design. Another source of errors can be caused by people who are still becoming familiar with using the technology. Where the accuracy of the system is to be measured when it is in full operation, it may be necessary to ensure the test subjects receive a practice run and appropriate training before the test commences. This is known as *habituation*.

5.3.3.2 After Data Collection

Once the data has been collected and matched, mistakes can often be spotted in the ground truth by looking at the match outliers. This is because a genuine match labeled as an impostor will typically have an unusually high score. Similarly, an impostor match labeled as genuine will tend to have an unusually low score. Before finalizing and testing results, a sanity check of the outlier results is highly recommend. Techniques such as the zoo plot (see Sect. 8.2.3) and the other techniques described in Chaps. 8 and 9 can also be used to spot labeling errors.

5.3.4 Data Collection

The collection of biometric samples is not only difficult because of the size of the population sample required, but there are a number of other challenges that need to be considered:

- Ensuring test subjects interact correctly with the acquisition system
- Precisely recording quality issues, including all failures to enroll and acquire
- Controlling or measuring the relevant environmental parameters
- Labeling the ground truth data correctly, including dates, times and person demographics
- Handling the distinction between presentations, attempts and transactions. An individual authentication attempt may fail, but the complete transaction can still be successful. See Fig. 5.3.
- Obtaining informed consent from test subjects, and handling the data securely and with privacy in mind
- Separating systems and data so that no artificial enhancement of scores can occur, either accidentally, through normalization or adaption, or deliberately
- Storing and managing all the data acquired in a format that allows easy versioning and reporting

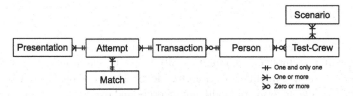

Fig. 5.3 Relationships between the different test elements in a scenario evaluation. First a biometric **presentation** is made, if there is not a failure to acquire, this results in an **attempt**. A number of attempts resulting in success or failure, depending on the number of tries allowed by the business rules, are grouped into a **transaction**. Each transaction belongs to a particular **person**, and each person may be part of zero or more **test-crews**. Finally, depending on the test requirements. each test-crew may be part of one or more **scenarios**.

5.3.5 Matching Issues

The general rule of biometric testing is that the more matching information available the better. However, large test sets can generate a prodigious amount of matching data. For instance, for a test with a million enrolled individuals in the gallery and one million probe images, the number of potential cross match results is one thousand billion (10^{12}). Managing large result sets can be challenging, and there are three main strategies for dealing with this volume of data: storing the top results, which is often the default setting for engines but makes analysis difficult since the score range has an arbitrary cut-off based on rank; storing the impostors and genuine matches above a threshold, which allows for accurate analysis of results for thresholds higher than the cutoff; and storing only results in the target range, which is useful when the threshold values of interest are known to lie within a particular range.

There are two issues relating to the algorithm configuration that can bias match results compared to operational performance. These are score normalization and template adaptation.

5.3.5.1 Normalization

It is possible to improve matching performance on smaller test sets using the knowledge of the enrollment database. This is called *score normalization*. This is discussed in more detail in Sect. 7.2.1.2. The effect of score normalization in testing is to artificially boost performance on small test sets compared to how the algorithm would perform using larger databases, or when tested with non-enrolled individuals. At least some impostor tests should be conducted using non-enrolled individuals to ensure that algorithms are not relying too heavily on score normalization. The average impostor score ranges when using un-enrolled people should be equivalent, within the margin of errors, to the average impostor scores on the enrolled set.

5.3.5.2 Adaptation

Some algorithms use *template adaptation* to improve the performance of the algorithm over time. Template adaptation usually works by enrolling verification samples when the sample quality is over a given level. The advantage of this technique is that changes due to aging or health are incorporated over time.

Enrollment is usually a carefully controlled process to ensure the enrolled template is of high quality. The disadvantage of template adaptation is that it can incorporate transient changes into the template (such as a cut on a finger, or a sore throat) leading to degraded performance overall. In the context of testing, template adaptation can mean that each time a test is conducted the individual's template changes. With a large sample of impostor tests, this might result in unwanted modifications, or it may artificially enhance results when the same enrollee is used in later testing.

Testing is best conducted without adaptation, since its effects are dependent on the testing sequence. Subsequent tests that explicitly explore the effects of adaptation on performance are recommended where this feature is to be used.

5.4 Reporting

Its important to consider the audience when performance reporting, as there are a wide variety of ways to show testing results. What is useful for algorithm developers and researchers will often be confusing or misleading to those making business decisions.

Reports should start with a statement about explicit assumptions used for the testing setup. This includes the mix of demographic or environmental conditions found in the target environment, and a demonstration how the test crew or test set accommodates these assumptions. A table should be included to explicitly indicate the number of test samples for each population group.

In many cases resource limitations will mean that not every possible situation was tested, so un-tested cases should be listed. The test environment is also important to document, including any external factors that might influence collection, such as ambient light, temperature and humidity, and the instructions for the test crew. The exact versions of software and any relevant setting information, such as thresholds or quality parameters, are required, particularly in situations where an updated engine(s) may tested in the future.

Information about the length of time between enrollment and verification and the frequency of visits by members of the test crew is vital information. Obtaining sufficient time separation, or longitudinal sampling, between enrollment and test transactions can be one of the more difficult tasks since is by nature time-consuming. If tests try to short-cut this requirement, generating test sets where the time separation is quite small (perhaps even only hours or days) the test results will tend to be significantly more positive than would be experienced in operation.

Biometrics For Business Cases

For non-academically minded managers, performance graphs, such as ROC and alarm curves, may be more confusing than helpful in the report body. In this case, bar charts showing the false reject rate for a fixed false accept rate are useful. Ideally, these graphs should also have error bars showing confidence levels. More technical graphs should be included in an appendix to the report to allow cross checking, and aid those seeking more detailed statistics.

Comparative reports need to emphasize the difference between the algorithms tested. For example, by looking for particular strengths or weaknesses of the algorithms under examination. The techniques for establishing individual and group level performance are covered in Chaps. 9 and 8, and are particularly relevant in this context.

Operational reports are likely to be focused on threshold and parameter settings, and on those individuals causing a disproportionate number of false accepts or rejects. An example report is shown in Fig. 5.4. The ISO Biometric Performance Testing and Reporting standard [3] specifies suggested elements for test reports. These items include most of the standard terms described in Chap. 7, but also include three different generalized false accept rates that take into account different ways of reporting the impact of failures to enroll and failures to acquire.

5.5 Conclusion

Ensuring accurate testing requires the planning and consideration of a wide variety of different factors. This chapter has explained these conditions to ensure that the testing process produces the desired answers about the expect performance of a real world biometric solution.

It is recommended that those embarking on the testing for the first time learn from the range of publicly available test reports and the ISO guidelines, and consider using dedicated software tools to assist them with this process. It cannot be over emphasized the importance of the data collection process and ground truth resolution, as this is often the most expensive part of the evaluation, and if it is not undertaken correctly the testing may be wasted. See Chap. 11 for a discussion on the impact of proper ground truth for surveillance systems.

Testing is likely to move from being a task that is primarily undertaken at the start of a project, to being a continuous and on-going process. This operational testing will be a vital part of future implementations for providing continual risk assessment, audits and on-going management.

Session Name: Algorithm 2

Description: Fingerprint Algorithm version 1.02a tested on
FVC2000 data-set[5]
Test Date: 9-June-2007
Session type: Verification

Number of users (people presenting a biometric): 100
Number of test subjects (people enrolled)   : 100

Number of samples : 800
Number of templates: 800
Number of enrollment failures: 0

Number of genuine matches : 5600
Number of impostor matches: 9900

Match threshold: 37.100
False non-match rate: 52/5600 = 0.929%, **95% CI:** [0.39% - 1.50%]
False match rate   : 90/9900 = 0.909%, **95% CI:** [0.68% - 1.17%]

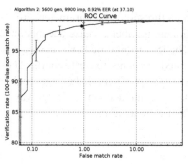 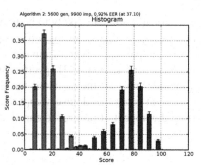

Failures to enroll: 0/800 = 0.00%
Failures to acquire: 0/800 = 0.00%

False reject rate: 3.71%

(FRR = FTA + FNMR*(1-FTA))

False accept rate: 3.11%

(FAR = FMR*(1-FTA))

Generalized false reject rate: 3.71%

(GFRR = FTE + (1-FTE)*FTA + (1-FTE)*(1-FTA)*FNMR)

Generalized false accept rate 1: 3.11%

(GFAR = FMR*(1-FTA)*(1-FTE)*(1-FTE))

Generalized false accept rate 2: 3.11%

(GFAR = FMR*(1-FTA)*(1-FTE))

Top 5 Individual False Matches		Top 5 Individual False Non Matches	
Person ID	Num False Matches	Person ID	Num False Non Matches
6	20	44	10
51	10	56	8
13	8	58	4
68	7	78	4
95	7	100	4

Fig. 5.4 Example performance report . Generated using Performix [1]

References

[1] Performix biometric research and analysis software. http://www.biometix.com/performix.htm (2007)

[2] Cappelli, R., Erol, A., Maio, D., Maltoni, D.: Synthetic fingerprint-image generation. Pattern Recognition, 2000. Proceedings. 15th International Conference on **3** (2000)

[3] ISO: Information technology – biometric performance testing and reporting – part 1: Principles and framework (ISO/IEC 19795-1:2006) (2006)

[4] Jovanovic, B., Levy, P.: A look at the rule of three. The American Statistician **51**(2), 137–139 (1997)

[5] Maio, D., Maltoni, D., Cappelli, R., Wayman, J.L., Jain, A.K.: FVC2000: Fingerprint verification competition. IEEE Trans. Pattern Anal. Mach. Intell. **24**(3), 402–412 (2002)

[6] Mansfield, A., Wayman, J.: Best practices in testing and reporting performance of biometric devices. NPL Report CMSC **14**(02) (2002)

[7] Mansfield, T., Kelly, G., Chandler, D., Kane, J.: Biometric product testing final report. Contract **92**, 4009,309 (2001)

[8] O'Toole, A.J., Phillips, P.J., Jiang, F., Ayyad, J., Penard, N., Abdi, H.: Face recognition algorithms surpass humans matching faces over changes in illumination. In: IEEE Transactions on Pattern and Machine Intelligence, vol. 29, pp. 1642–1646 (2007)

[9] Phillips, P.J., Scruggs, W.T., O'Toole, A.J., Flynn, P.J., Bowyer, K.W., Schott, C.L., Sharpe, M.: FRVT 2006 and ice 2006 large-scale results. Tech. Rep. NISTIR 7408, National Institute of Standards and Technology (????)

[10] Reynolds, D.A., Doddington, G.R., Przybocki, M.A., Martin, A.F.: The NIST speaker recognition evaluation - overview methodology, systems, results, perspective. Speech Commun. **31**(2-3), 225–254 (2000)

[11] Schuckers, M.: Estimation and sample size calculations for correlated binary error rates of biometric identification rates. Proceedings of the American Statistical Association: Biometrics Section (2003)

Chapter 6
Definitions

This chapter introduces the terms that will be used throughout the rest of the book, which is a more formal treatment of many of the subjects introduced in the preceding chapters.

There is a considerable amount of inconsistency among the terminology currently used within the biometric research and industrial communities. This has been due in part to a degree of isolation in which the fields for different modalities have developed. For example, the terminology and graphs used by speech verification researchers can differ considerably from those used within the face recognition community. This leads to unnecessary confusion between people within the field, and more importantly, between the field as a whole and the general public.

The best effort to date to establish standard terminology is ISO/IEC 19795-1, *Information technology - Biometric performance testing and reporting.* This document grew out of previous work in the UK Biometric Working Group's *Best Practices in Test and Reporting Performance of Biometric Devices.* In an effort to help further establish the standards, our definitions endeavor to be consistent with these documents.

6.1 General

Biometrics Biometrics is the automatic identification of an individual based on his or her physiological or behavioral characteristics.

Biometric A measure of a biological or behavioral characteristic used for recognition. There are four requirements for a biometric: every person must have it, it should be sufficiently different for every person, it should remain constant over time, and must be able to be measured quantitatively.

Characteristic In some cases there will be multiple instances of the same biometric from one person. For example, most people have two eyes (for iris recognition) and ten fingers (for fingerprint recognition). Each instance is a characteristic. Unless otherwise specified, when talking about matching the two different samples of a biometric from the same person, it is assumed that the same characteristic is being matched.

Biometric system The authentication system in its entirety. In particular, it consists of the integrated hardware and software used to gather samples, conduct biometric verifications or identifications, and return or log the results.

Modality   A mode or type of biometric used for recognition. For example, three common biometric modalities are face, fingerprint and voice.

Multibiometrics A multi-biometric systems, or multimodal biometric system, uses more than one biometric characteristic for authentication. This can be multiple instances of the same modality, such as two fingers or both eyes, or two different modalities, such as face and fingerprint. There two major advantages over using a single biometric. Firstly, people with damage to one biometric may still be able to use the system using the other. Secondly, as the individual biometrics have a high degree of statistical independence, there are significant performance benefits that can be gained through multimodal fusion.

6.2 Biometric Data

Sample   An instance of a user's biometric, as captured by the system's sensor(s). Samples are unprocessed digital data, such as a digital image of a user's face, a sound recording of their voice, or a video clip of them walking. A sample may be converted to a template (enrollment) or matched against an existing templates (verification or identification).

Feature   Distinctive information extracted from a biometric sample, and stored electronically. Features are used to distinguish samples, and should be robust to changes in the environment and aging. For example, the most common features for fingerprint recognition are the types and locations of all minutiae points (locations where fingerprint ridges split or terminate).

Template (Enrollment) A template is a compact, electronic representation of a biometric sample that is created at the time of enrollment, and stored in the system database for future reference and matching. There are two advantages to storing templates rather than the original sample. Firstly, the features have already been extracted, alleviating the need to perform this computationally expensive step more than once. Secondly, in theory the templates embody only the distinctive information of the biometric, and discard the background, noise, and other information. This leads to much smaller storage requirements than one would need to store the original sample.

Match (Verification) A comparison between two biometrics, typically a verification sample and an enrollment template. The output of a match is a match score. For example, a match between an image of person A's face and an image of person B's face may lead to a score of 32.

Genuine match A match between two instances of the same biometric characteristic from the same person.

Impostor match A match between two different biometric characteristics. This is usually a match between two different people, but also includes a match between two different characteristics of the same person, such as matching the left iris against the right iris.

Match score A single numerical value that represents the degree of similarity between two samples. A high match score indicates a high degree of similarity, and a low match score indicates dissimilarity. Match scores are used as the basis for verification and identification decisions.

Score threshold A parameter of a biometric system for determining match decisions. A match that achieves a score exceeding the score threshold is labeled a match, otherwise it is labeled a non-match.

Match decision Accepting or rejecting an input match. The decision is based on the match score and its relation to the score threshold. For example, for a score threshold of 45, a match with a score of 50 will be accepted, and a match with a score of -2 will be rejected. A match decision may be correct or incorrect, depending on the decision and whether or not the match is genuine or impostor.

Ground truth Ground truth is the actual state of nature, as pertaining to a match. In other words, it is the truth of whether or not a given match is genuine or impostor.

Probe   For an identification system, the input sample is sometimes referred to as the probe. This terminology is mostly used for face recognition identification systems.

Gallery   For an identification system, the enrollment database is sometimes referred to as the gallery.

Watchlist   The list of people who a system is designed to detect (i.e. the enrollments).

Candidate list For an identification system, the set of templates returned as potential matches is known as the candidate list. Candidate lists can be selected based on rank (e.g. the top 5 ranked results) or score (e.g. all matches with a score ≥ 55).

6.3 Biometric Systems

6.3.1 People

User   A person who presents a sample to a biometric system for verification or identification, regardless of whether or not they are enrolled.

Test subject A person who has been recruited to participate in a biometric evaluation.

Test crew   The set of test subjects enrolled during an evaluation test

Population   The set of all users who would be eligible to use a live system. For a biometric evaluation, test subjects should be a representative sample of the target population.

Operator   A person who has a role in the operation of a biometric system. For example, in a surveillance system an operator may visually examine all images on the candidate list to confirm the match.

6.3.2 User Interaction with a Biometric System

Enrollment The process of enrolling a user into a biometric system. This generally involves creating a template for the user and storing it in the database.

Presentation The submission of a biometric sample to the system, which will be used for either enrollment, verification, or identification.

Transaction A sequence of presentations and attempts, culminating in an enrollment, verification, or identification. For example, the first presentation may fail due to insufficient quality. However, if the second presentation is successful, the overall transaction may be successful.

Verification The submission of a biometric sample and a claim of identity with the aim of authentication. For example, consider a biometrically enabled ATM. In this case, a person's bank card would contain their claim of identity, and a verification would be conducted to ensure that the card holder is the correct owner.

Identification The submission of a biometric sample with the aim of determining if the user is enrolled in the database, and if so, finding their identity. For example, consider an Automated Fingerprint Identification System (AFIS). A latent fingerprint is found at the scene of a crime, and an identification is conducted to determine the identity of the person who left the fingerprint.

6.3.3 System Types

Verification system A verification system takes as input a sample and a claim of identity. The system matches the sample against the enrollment for the claimed identity, and evaluates the validity of the claim. The output is a verification decision, either accepting or rejecting the claim.

Identification system An identification system takes as input a sample, and returns the most probable identity, or set of identities, from the enrollment database.

Open-set identification system This is a special case of an identification system
        where some users are not enrolled in the database. Open-set identifi-
        cation systems are the most common real world identification system
        as it is generally difficult to enforce that all people who can potentially
        use a system are enrolled. An example of an open-set system is a bio-
        metric surveillance system that monitors a public area in an attempt to
        detect persons of interest.

Closed-set identification An identification system where is it known a priori that all
        potential users are enrolled. Closed-set identification is often used for
        system testing and evaluation, but is rare for real-world applications.

6.3.4 Applications

Watchlist Many biometric systems fall under the description of a watchlist sys-
        tem, which enrolls people of interest, and identifications are conducted
        to check for matches. A common example would be an AFIS fingerprint
        system, where latent fingerprints are matched against all enrollments in
        an attempt to identify the unknown person.

Surveillance system Surveillance systems capture biometrics remotely using surveil-
        lance sensors, such as high-resolution video cameras. Some surveil-
        lance systems are designed to be covert, in which case users are un-
        aware of the existence of the system. The system matches the biometric
        samples against a watchlist in order to automatically identify persons
        of interest.

Access control system An access control system uses biometric authentication to
        control access to a restricted area. For example, an access control sys-
        tem at a local gym might use fingerprint verification to ensure that only
        members can use the facilities.

Identity document systems An identity document system maintains records of iden-
        tity for a population. Examples include the databases maintained by
        driver's license and passport authorities. Biometrics can be used im-
        prove the integrity of the proof of identity (POI) process.

Negative Identification Some identification systems are designed to ensure that in-
        put samples do *not* belong to anyone already enrolled. An example
        would be a social welfare system that uses fingerprint matching to make
        sure people do not claim benefits under multiple identities.

Time and Attendance Many companies use systems that automatically log when
        an employee arrives at work, and when they leave. Biometrics can be
        effectively used to avoid situations where people cheat the system by
        getting fellow employees to log them in or out.

6.4 Biometric Evaluation

6.4.1 Evaluation Types

Technology evaluation A technology evaluation is focused on the performance of
the biometric matching algorithm. The tests are run in an offline fashion, using a set of samples selected expressly for the evaluation. For
example, the Fingerprint Verification Competition (FVC) and Face
Recognition Grand Challenge (FRGC) are well known technology evaluations designed to establish the state-of-the-art in fingerprint and face
recognition performance.

Scenario evaluation For a scenario evaluation, the system as a whole is tested in
an environment that simulates, as closely as practical, the intended operational environment. For example, for a surveillance trial this may
include building a mock environment that resembles a likely site location, and having subjects walk through the system to test the detection
accuracy and false alarm rate.

Operational evaluation For an operational evaluation, the system is tested in the
actual application environment, using members of the target population for data collection. For example, testing may be conducted on an
existing speech recognition system to determine its accuracy and vulnerabilities. Determining ground truth and generating impostor results
are challenges for operational evaluations.

6.4.2 Performance Measures

False non-match A genuine match that is declared to be a non-match. This is due
to the match receiving a score below the match threshold.

False match An impostor match that declared is declared to be a match. This is due
to the match receiving a match score above the match threshold.

False non-match rate (FNMR) The proportion of genuine matches that are falsely
non-matched.

False match rate (FMR) The proportion of impostor matches that are falsely matched.

False reject rate (FRR) The proportion of genuine transactions that are rejected by
the system.

False accept rate (FAR) The proportion of impostor transactions that are accepted
by the system.

Equal error rate (EER) The error rate at which the false accept rate equals the false
reject rate. The EER can be used to summarize the performance of a
system, as it contains both false match and false non-match information.

Identification rate The proportion of genuine identification attempts for which the correct enrollment is returned in the candidate list.

False-negative identification error rate The proportion of identification attempts in which a user enrolled in a system does not have at least one of their templates returned among the candidate list.

False-positive identification error rate The proportion of identification attempt in which a non-empty candidate list is returned, which does not include any genuine matches.

Failure to enroll If the failure occurs during enrollment, it is known as a failure to enroll. The proportion of enrollment transactions that fail is known as the failure to enroll rate (FTE).

Failure to acquire If an error occurs while acquiring the biometric sample during a verification or identification, it is known as a failure to acquire. The proportion of verification or identification attempts that fail for this reason is the failure to acquire rate (FTA).

6.4.3 Graphs

Histogram A histogram is a non-parametric method of displaying genuine and impostor match score distributions.

ROC curve A receiver operating characteristic (ROC) curve summarizes system performance by plotting false match rate vs. verification rate pairs for a range of match thresholds.

DET curve A detection error trade-off (DET) curve summarizes system performance by plotting false match rate vs. false non-match rate pairs for a range of match thresholds.

CMC curve Cumulative match characteristic (CMC) curves summarize performance for closed-set identification systems. The y-axis is the identification rate, and the x-axis is the size of the candidate list, which contains the top X ranked results.

Alarm curve (Watchlist ROC) Performance for open-set identification systems can be displayed using alarm curves, which plot the detection rate against the false alarm rate over a range of thresholds.

Zoo plot A graph that shows individual user performance. The position of each user is determined by the genuine matching performance (i.e. their ability to authenticate themselves) and their impostor matching performance (i.e. their likelihood of contributing to false matches).

6.4.4 The Biometric Menagerie

Sheep   Sheep make up the majority of the population of a biometric system. On average, they tend to match well against themselves, and poorly against others. Their genuine and impostor match score distributions are consistent with the system wide distributions.

Goats   Goats are subjects who are difficult to verify. They are characterized by a low genuine match score distribution, and may be involved in false rejects.

Lambs   Lambs are vulnerable to impersonation. When matched against by others, the result is relatively high match scores (i.e. high impostor distribution), leading to potential false accepts.

Wolves   Wolves are successful at impersonation, and prey upon lambs. When matched against enrolled users, they receive high match scores. In some systems wolves may cause a disproportionate number of the system's false accepts.

Doves   Doves are the best possible users of a biometric system. They match exceedingly well against themselves, and are rarely mistaken as others.

Worms   Worms are the worst conceivable users of a biometric system, and are characterized by low genuine scores and high impostor scores. If present, worms are responsible for a disproportionate number of system errors.

Chameleons Chameleons always appear like others, receiving high scores for matches against themselves and others.

Phantoms   Phantoms never appear similar to anyone, including themselves. Phantoms have low genuine and impostor score distributions.

6.4.5 Vulnerability and Security Definitions

Artifact   An object that has been created with the intention to be used as a fake biometric in order to attack a system. An example is a latex fingerprint.

Spoofing   The act of using an artifact to attempt to fool a biometric system.

Liveness detection The determination that a biometric presentation is from a live human rather than an artifact. This could be by detecting properties of the biometric presentation that relate specifically to active biological processes such as pulse rate or heat.

Threat   A possible technique that might be used to breach the system security (also known as an attack). For instance, the use of an artifact is a type of biometric threat.

Attacker   The person that is attempting to undertake an attack. The attacker is the person that is actually attempting a fraudulent enrollment or authentication, rather than a malicious hacker facilitating an attack through the IT infrastructure.

Mitigation Techniques that are used to address the risk of a specific attack. An example of a mitigation might be the use of liveness detection.

Residual risk The remaining risk after a mitigation has been applied to a given threat. For example, if body heat was used for liveness detection, a residual risk would be that an attacker could warm up the artifact before use.

Attacker strength The characterization of the resources and capability of the attacker. This includes their technical ability and knowledge, the level of system access, and the amount of time and resources available to them.

Vulnerability risk The chance that a motivated attacker will be able to break a system, for a given threat and attacker strength.

Proof of Identity The set of identity factors that are used to establish trust in an identity. This commonly includes documents such passports and birth-certificates.

Biometric theft Fraud undertaken without the knowledge of the person whose biometrics are being faked. This requires the attacker to have access to either covertly obtained biometric samples, stolen biometric templates or other mechanisms to fraudlently obtain an individual's identity.

Cooperative fraud This is fraud that uses cooperation from the person to be mimicked as a mechanism for them to allow others to use their identity. This may, for instance, be to allow others to travel under one's issued identity document.

6.4.6 Statistical Terms

Null hypothesis (H_0): The claimed identity is true. In other words, the two biometric samples have been obtained from the same characteristic of the same person.

Alternative hypothesis (H_1): The claimed identity is false. In other words, the sample does not match the enrollment template.

False non-match (Type I error) A Type I error occurs when the null hypothesis is true, but is rejected nonetheless. This is also known as an α error. The result of a Type I error is falsely rejecting a genuine match, and is therefore known as a false non-match.

False match (Type II error) A Type II error occurs when the null hypothesis is false, but is accepted. This is sometimes referred to as a β error. The result of a Type II error is falsely accepting an impostor match as genuine, thus resulting in a false match.

Sampling error An error caused by sampling a finite subset of a population, and using the result as representative of the whole user population.

Systematic error A bias in a measurement (performance rate) that leads to the result being consistently too high or too low. An example would be a faulty sensor, or a over-represented population demographic among the test subjects. Systematic errors can be kept to a minimum through careful planning and control, but they are practically impossible to eliminate completely.

Variance   The variance of a probability distribution is a measure of its dispersion. When the variance is low, an observed result is likely to be close to the real value.

Confidence interval An interval estimate of an error rate, along with an associated probability that the interval contains the true result. For example, an EER may be reported as "The EER is 0.23%, with a 95% confidence interval [0.19%, 0.41%]". Wide confidence intervals reflect a high degree of uncertainty in a results, while narrow intervals indicate a reliable result.

Statistically significant A result that is unlikely to be due to chance. In theory, only statistically significant results should be reported. For example, one might say "adjusting the lighting conditions in the capture environment resulted in a statistically significant reduction of false rejections".

Part II
The Biometric Performance Hierarchy

In Part I of this book, a high-level view of biometrics and biometric evaluations was given. Part II examines many of these concepts again in more detail. In particular, the question "Given a biometric system, how do we determine how well it is performing?" is addressed. This is a broad question, and many approaches have been developed over the years. The actual answer depends on what particular aspect of the system you are interested in. For example, you may be interested primarily in usability. In this case, you would elicit subjective feedback from a set of test subjects. Other types of performance measures may include throughput rates, operator experience, performance scaling, cost analysis, etc. The focus of the next three chapters is on the quantitative analysis of biometric performance.

Conceptually, a biometric system can be thought of as a hierarchy, with individual people at the bottom, groups of people in the middle and the system as a whole at the top. At each level, different approaches to performance evaluation are appropriate. Traditionally, the focus of evaluation has been on system-level performance, as outlined in Chapter 7. However, anyone who has real-world experience with biometric systems will have observed that there is an imbalance in the performance of individual users. In particular, if often seems that a small set of people are responsible for most of the errors. The *biometric menagerie* is the established method for characterizing users. This approach is explained in Chapter 8, and the theory is extended with novel insights into the role of individualized error rates. Since individual performance can vary, one may expect to observe trends in the user population. Data mining is a technique for automatically discovering knowledge in data, and there is little existing information about its application to biometric data. This group-level analysis is at the middle of our performance hierarchy, and is the subject of Chapter 9.

Chapter 7
System Evaluation: The Statistical Basis of Biometric Systems

Some biometric systems perform well, and others perform poorly. This chapter examines in detail the methods for establishing and reporting system accuracy. The manner in which a system is evaluated depends on how it operates. There are two fundamental types of biometric systems: verification systems and identification systems. Verification systems are the subject of Sect. 7.1. The section begins with a review of statistical theory, which is at the heart of the biometric matching process. After these foundations have been laid, the discussion proceeds to develop the rates and graphs used for reporting verification system performance. This is followed by an examination of identification systems in Sect. 7.2. This section differentiates between verification and identification systems, and demonstrates how identification is actually composed of a series of verifications. Section 7.3 examines the role of chance in biometric analysis. As will be seen, the reliability of performance testing actually depends on the size of the experiment. Finally, in Sect. 7.4 other quantitative measures of system performance are explored.

By the end of this chapter, you should have a firm understanding of:

- The statistical basis of verification decisions (Sect. 7.1.1).
- The difference between verification and identification systems. (Sect. 7.2.1).
- The common performance measures and graphs used for verification (Sect. 7.1) and identification (Sect. 7.2) systems, and when they are applicable.
- The dependence of identification performance on the number of enrollments in a system (Sect. 7.2.1.1).
- The role of statistical uncertainty in biometric matching, and its impact on the way performance is measured and evaluations are conducted (Sect. 7.3).
- Enrollment and acquisition errors, and how they impact system performance (Sect. 7.4.1).

7.1 Verification

The key feature of verification systems is that they involve a *claim of identity*. For example, consider a biometrically enabled ATM. The customer's bank card intrinsically contains their claim of identity, and they present their biometric (e.g. finger) to verify that they are the rightful owner of the bank card. There are two potential errors that can occur: the ATM may incorrectly reject an attempt by the rightful owner, or it may falsely accept an attempt by an impostor. This section outlines the statistical underpinnings of verification systems. Firstly, the problem is framed in terms of a statistical method known as hypothesis testing. This places biometric verification on a firm, statistical basis. Secondly, the performance statistics and graphs which summarize the performance of verification systems are presented.

7.1.1 Hypothesis Testing *

Verification systems make a discrete decision, accept or reject, based on probabilistic data. The specific question addressed in this section is "How likely is the verification system to make an incorrect decision?" The phrasing of the question implies a *likelihood* associated with errors. In other words, errors do not occur deterministically, but rather in a probabilistic manner. Statistical theory is the field that is used to analyze and interpret probabilistic data, and is therefore the primary tool for the evaluation of biometric systems. *Hypothesis testing* is an area of statistics that deals with making decisions based on uncertain information, and examines the inherent risks involved. The approach is well suited to biometric verification, and places it within a robust statistical framework.

7.1.1.1 Problem Formulation

Hypothesis testing begins by formulating a *null hypothesis* (H_0), which is presumed true unless there is evidence to the contrary. If there is sufficient evidence to the contrary of the H_0, an alternative hypothesis H_1 is accepted. In the case of biometric verification, the hypotheses refer to the user's claim of identity, and are formulated as follows:

Null hypothesis (H_0): The claimed identity is true (genuine). In other words, the two biometric samples have been obtained from the same characteristic of the same person.

Alternative hypothesis (H_1): The claimed identity is false (impostor). In other words, the sample does not match the enrollment template.

It is the goal of the verification system to use the information available to either accept or reject H_0. This information, typically a scalar value, summarizes all the relevant information, and is known as the *test statistic*. In the case of biometric systems, the test statistic is the match score, which is a measure of similarity between two biometric samples.

The distribution of the test statistic when H_0 is true (i.e. the match is genuine) differs from the distribution when H_1 is true (i.e. the match is an impostor). In general, the match scores for impostor matches will be lower than the match scores for genuine matches.[1] It is the distributions of the test statistics, and particularly the overlap between them, that determines the probabilities of verification errors. Example probability density functions of the two distributions are illustrated in Fig. 7.1.

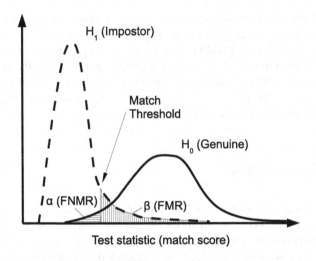

Fig. 7.1 Probability density functions of the test statistic for H_0 and H_1. H_0 represents genuine match similarity scores, and H_1 is impostor match similarity scores. At a given match threshold, the area under H_0 to the left represents the probability of false non-matches. The area under H_1 to the right of the match threshold represents the probability of false matches.

When the match score is high, this is evidence in favor of H_0. However, a low match score is evidence against H_0. The *match threshold* is used to establish the point below which a score is too low support the identity claim, and thereby reject the null hypothesis. Given a match score s and a match threshold t:

$$\text{Verification decision} = \begin{cases} \text{Accept } H_0 & s \geq t \\ \text{Reject } H_0 & s < t \end{cases}$$

---

[1] This assumes that the match score is a measure of similarity. For some systems the match score may be a distance, or dissimilarity, measure. In this case, the match score distribution for genuine matches will be lower than for impostor matches.

7.1.1.2 Decision Errors

There are two possible errors that can occur:

Type I error A Type I error occurs when the null hypothesis is true, but is rejected
nonetheless. This is also known as an α error. For biometric verification, the result of a Type I error is falsely rejecting a genuine match,
and is therefore known as a *false non-match*.[2]

Type II error A Type II error occurs when the null hypothesis is false, but is accepted. This is sometimes referred to as a β error. For biometric verification, the result of a Type II error is falsely accepting an impostor
match as genuine, thus resulting in a *false match*.

The probabilities of false match and false non-match errors can be computed from
the probability density functions for H_0 and H_1. The probability of a false non-match
is the false non-match rate (FNMR), which is the area under the H_0 distribution to
the left of the match threshold. This area is indicated in Fig. 7.1 as the area shaded
with horizontal lines. The probability of a false match is the false match rate (FMR),
which is the area under the H_1 distribution, to the right of the match threshold. In
Fig. 7.1, this is the area shaded with vertical lines.

7.1.2 Performance Rates

The previous section presented a method for making verification decisions (accept
or reject) based probability distributions for the null hypothesis (i.e. the user is
telling the truth) and the alternative hypothesis (i.e. the user is lying). The determination of these prior probabilities is now examined, and in particular how they
are used to compute the standard performance measures.

7.1.2.1 Representing Match Score Distributions

Traditional hypothesis testing typically uses a family of test statistics whose distributions are known a priori (such as the t-test or z-test). However, in the case of
biometric systems, the distributions of the genuine and impostor test statistics must
be determined empirically. In other words, the distributions are estimated by computing the test statistic for a large sample of genuine and impostor matches.

---

[2] In statistics literature, a Type I error is often referred to as a *false positive*, since the alternative
hypothesis is falsely accepted. This terminology may cause confusion in the case of biometrics,
because the outcome of *falsely accepting* the alternative hypothesis (i.e. that the samples belong to
different people) is *falsely rejecting* a genuine match. The confusion arises from the conflicting use
of the term "acceptance", which is applied to the claim of identity in one case, and the alternative
hypothesis in the other. The convention used here is that "acceptance" refers to the claim of identity
[15].

Collecting sample match scores is a vital step for system evaluation, and it is important that steps are taken to ensure robust data collection. The following are some important caveats for sample score collection (also see Chap. 5):

- The subjects (and their templates) used for generating the sample match scores should be representative of the target population.
- Scores should be sampled from the full range of possible values. For example, some biometric engines only output the top ranked scores by default in the matching logs. Without samples from the entire range of scores, it is not possible to calculate verification error rates.
- As many genuine and impostor samples as practical should be collected in order to maximize the precision of the estimates (see Sect. 7.3).
- It is very important *not* to assume a parametric form of the match score distribution, such as normality. Such assumptions are rarely justified, and may severely bias error rate computations.

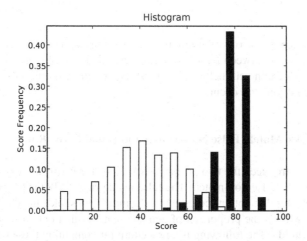

Fig. 7.2 A match score histogram. The black bars represent genuine scores, and the white bars are impostors.

After a sample of genuine and impostor scores have been collected, they are used to estimate the match score distributions. Note that a finite number of sample scores are used to estimate the true (unknown) distributions, which are continuous. *Histograms* tabulate the number of samples in disjoint ranges of scores (bins), and are commonly used in statistics to graphically represent discrete distributions. Figure 7.2 contains an example of a match score histogram.

The distance between, or overlap of, the genuine and impostor match score distributions has an important interpretation for biometric data (see Fig. 7.3). Generally, the greater the distance between these distributions, the lower the system's error rates. This can be demonstrated by considering an extreme example where there is

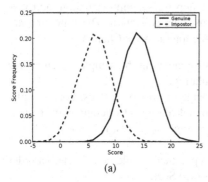 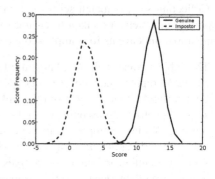

(a)

Fig. 7.3 The relationship between score distributions and accuracy. (a) The overlap between the genuine and impostor score distributions is large, therefore there is a range of scores (8-12) where both genuine and impostor scores are likely. This will lead to false match or false non-match errors, depending on the match threshold. (b) There is little overlap between the genuine and impostor score distributions. If the match threshold is set to 7.5, false match and false non-match errors will be rare.

no overlap between the two distribution. In this case, one can set the match threshold anywhere in the area between the distributions. At this threshold, all scores above the threshold are genuine, all below are impostors, and therefore no false match or false non-match errors will occur.

7.1.2.2 The False Match, False Non-Match and Equal Error Rates

Section 7.1.1.2 introduced the two primary error types for verification systems: the false match and the false non-match. The false match rate (FMR) is the proportion of impostor matches that obtain scores above the match threshold, and the false non-match rate (FNMR) is the proportion of genuine matches that obtain scores below the match threshold.[3] The following is a procedure for computing these rates:

1. Select a match score threshold t
2. Sort the genuine (impostor) match score in increasing order
3. Divide the number of genuine (impostor) scores $< t$ ($\geq t$) by the total number of genuine (impostor) scores to find the false non-match (false match) rate

Note that the FMR and FNMR are functions of t. Denote FNMR(t) and FMR(t) to be the corresponding rates at the given threshold.

---

[3] Recall that a false match is a Type II error. The *power* of a hypothesis test is the probability of rejecting a false null hypothesis. Intuitively, this is the probability of correctly non-matching an impostor (1-FMR). The Type I error (false non-match) rate of a system can be reduced to an arbitrarily small level by lowering the match threshold. However, for a system to have statistical power, it must still be able to reject impostors.

Generally, one would like to know how a system performs over a range of thresholds, so the above procedure is repeated many times for different values of t. The thresholds at which to compute the error rates can be chosen in a variety of ways. For example, they can be selected at equally spaced intervals throughout the range of possible match score values. Alternatively, one can compute the error rates for each unique score that has been observed.

The FMR(t) and FNMR(t) are plotted together in order to see their relation to each other. Figure 7.4 (b) contains these plots for the error rates corresponding to the score distributions of Fig. 7.4 (a). As the threshold value increases, the FMR varies from 100% (every impostor match is accepted) to 0% (every impostor match is reject). Conversely, the FNMR varies from 0% (every genuine match is accepted) to 100% (every genuine match is rejected) as the threshold increases.

The point where the FMR curve and FNMR curve intersect is known as the *equal error rate* (EER). It is called this because at this threshold, the FMR and FNMR are equal. With respect to the score distributions, the EER occurs at the threshold t where the area under (i.e. the integration of) the genuine distribution $< t$ equals the area under the impostor distribution $\geq t$. Note that this is generally not the same as the threshold at which p(s|Genuine)=p(s|Impostor) for a match score s (i.e. at the crossover of the probability density functions). This can be observed in Fig. 7.4 (a). The EER threshold is indicated by a vertical line, and does not correspond to the crossover of the probability density functions. However, the EER threshold does correspond to the crossover point of the FMR and FNMR curves in Fig. 7.4 (b).

Reporting the EER

The EER is often used as a simple way to summarize the performance of a biometric system. This due to its convenience as a single value with an intuitive interpretation. However, it should be kept in mind that there is no other special distinction of the value, and systems in production are rarely configured to operate at the EER. Furthermore, systems should not usually be compared based on this value alone. Simply because system A has a lower EER than system B, this does not necessarily imply that system A will outperform system B over all thresholds.

7.1.2.3 Selecting an EER Point *

The true (unknown) distributions for the genuine and impostor scores are continuous, yet are estimated using a finite number of samples. Therefore, the FMR and FNMR curves constructed are discrete. This raises difficulties for the computation of the EER, as an exact point of crossover may not exist (especially for small data sets). For example, in Fig. 7.4 (b) the FMR and FNMR curves appear smooth. However,

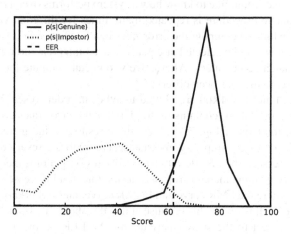

(a) Genuine and impostor probability distribution functions

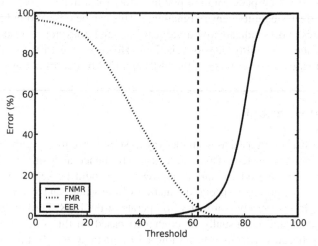

(b) The false match rate and false non-match rate for the distributions in (a)

Fig. 7.4 (a) Genuine and impostor distributions. (b) The corresponding FNMR and FMR over a range of thresholds. At the minimum threshold value, the FMR is always be 100% because all matches (including impostors) are accepted. At the maximum threshold value, the FNMR will always be 100% because all matches (including genuine) are rejected. In both graphs the threshold at the EER is indicated by a vertical dashed line.

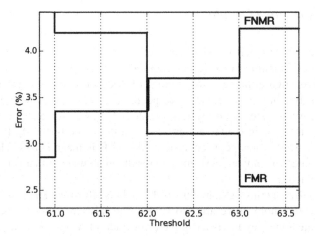

Fig. 7.5 The discrete nature of the crossover point between the FMR and FNMR curves of Fig. 7.4 (b). The value of the equal error rate appears to be around 3.5%, but a formal definition of EER is necessary to select a specific value. For this example, the EER is 3.41% using the formula from the FVC2000 competition.

Fig. 7.5 contains a zoomed view of the crossover region. From this view the discrete nature of the curves becomes apparent, and it is not self-evident what the correct value of the EER is because there is no threshold t at which FMR(t) = FNMR(t). At a threshold of 62, FMR>FNMR, while at 63 FMR < FNMR. There are several possible resolutions. One can interpolate the curves (e.g. using splines) to compute a hypothetical crossover point. However, this has the disadvantage that the precise crossover point depends not only on the data, but also on the interpolation method used. Another approach is to create a formal definition of the EER based on the false match and false non-match values in the region. For example, the FVC2000 competition uses the following estimation [14]:

$$t_1 = \max_t\{t|\text{FNMR}(t) \leq \text{FMR}(t)\},$$

$$t_2 = \min_t\{t|\text{FNMR}(t) \geq \text{FMR}(t)\},$$

$$[\text{EER}_{low}, \text{EER}_{high}] = \begin{cases} [\text{FNMR}(t_1), \text{FMR}(t_1)] & \text{if } \text{FNMR}(t_1) + \text{FMR}(t_1) \leq \\ & \qquad \text{FMR}(t_2) + \text{FNMR}(t_2) \\ [\text{FNMR}(t_2), \text{FMR}(t_2)] & \text{otherwise} \end{cases}$$

and the EER is $(\text{EER}_{low} + \text{EER}_{high})/2$ [14].

7.1.3 Performance Graphs

The FMR and FNMR performance measures specify error rates at a given match score threshold. This imposes several limitations. Firstly, when evaluating a system performance is often of interest over a range of thresholds. Secondly, the threshold values themselves are only relevant to a specific algorithm, and cannot be directly compared between systems. For example, system A may assign match scores in the range [0, 1], and system B may assign scores in the range [-10, 30]. In this case, directly comparing scores between systems A and B is meaningless. Therefore, a method is needed to compare performance results in an manner that is independent of specific match scores.

Figure 7.4 (b) contains a graph of the FMR and FNMR over a range of thresholds. At any given threshold, there are two values: the false match rate, and the false non-match rate. One way to visualize the error trade-off is by taking the FMR and FNMR and plotting the error rates over a range of thresholds. Two standard plots in the literature that do precisely this are the ROC curve and DET curve.

7.1.3.1 ROC Curves

A *receiver operating characteristic* (ROC) curve plots the false match rate against the verification rate. The verification rate is the likelihood of correctly accepting a genuine match, or 100-FNMR. Figure 7.6 contains the ROC curve corresponding to the data in Fig. 7.2. As illustrated in Fig. 7.6, the *x*-axis of an ROC graph is typically logarithmic since low false match values are of more interest, and a logarithmic scale better distinguishes values in this range. ROC curves with a logarithmic *y*-axis are sometimes found in the literature as well.

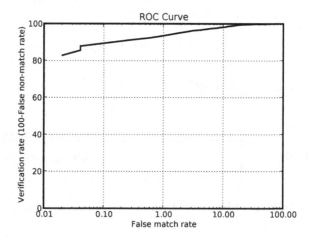

Fig. 7.6 An ROC curve. At a false match rate of 0.1%, the verification rate is around 90%.

7.1.3.2 DET Curves

A *detection error trade-off* (DET) curve is a variant of the ROC curve. The primary difference is that the y-axis is the false non-match rate, as opposed to the verification rate (see Fig. 7.7). In other words, both axes represent error rates. As with the ROC curve, the axes of a DET curve are often plotted using logarithmic scales. Alternatively, DET curves are sometimes plotted using a normal deviate scale [16]. In the case that the underlying distributions approximate normality, the DET curves plotted using a normal deviate scale will be roughly linear. This tends to make the graphs visually pleasing, and helps highlight the differences between similarly performing systems. However, note that the choice between ROC and DET curves, and the scale of the axes, is purely aesthetic, and all variants contain equivalent information.

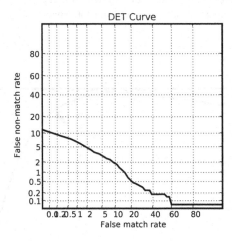

Fig. 7.7 A DET curve plotted using the normal deviate scale. At a false match rate of 2%, the false non-match rate is 5%.

7.1.3.3 Interpreting ROC Curves

ROC curves are widely used for reporting results and comparing classifiers in a number of fields, including image processing, data mining, machine learning, and pattern recognition. Due to its prevalence in the research community, their interpretation has been widely studied [8]. As mentioned in the previous section, the differences between ROC and DET curves are primarily aesthetic, with each emphasizing different aspects of system performance. The discussion of this section is focused on ROC curves, however the concepts are equally applicable to DET curves.

The fundamental idea behind ROC curves is that there is a trade-off that must be optimized between correct detections and false alarms. As the match threshold is lowered, both detections and false accepts increase. As the match threshold is raised, both detections and false alarms decrease. In biometric systems, these conflicting rates must be balanced to find an acceptable operating point, and each point on an ROC curve represents a possible combination.

The upper left corner of an ROC graph is of particular interest, as it represents a perfect biometric system. At this point, the verification rate is 100% and the false match rate is 0%. In other words, this biometric system makes no errors on the corresponding test set.[4] In practice, "perfect" performance is not achievable, due to the inherent uncertainty of biometric authentication. However, all systems strive to achieve curves near this corner. Therefore, when comparing the curves for two systems, the curve closest to the upper left corner represents superior performance. For example, observe the ROC curves for systems A and B in Fig. 7.8. At a false match rate of 0.01%, system A has a verification rate of around 90%, while system B's error rate is approximately 95%. Therefore, system B is the better system, verifying the heuristic that better systems are those whose ROC curves are closest to the upper left corner.

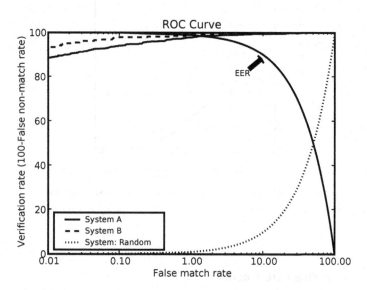

Fig. 7.8 ROC curves showing the performance of three hypothetical systems. A line is also indicated that intersects ROC curves at their equal error rate.

There are several other areas of an ROC graph that have interesting interpretations. Areas on the left of the graph represent "secure" operating points as this is where false match rates are low - the system is more likely to reject legitimate users than accept impostors. The upper right hand corner represents a policy of "accept

---

[4] For a DET curve, the bottom left corner represents a system with no errors.

everybody". In this case, verification rates are 100%, as everyone is accepted. However, false match rates are also 100%. In general, points in the upper-right region can be considered "convenient", as legitimate users will rarely be falsely rejected.

Figure 7.8 contains an ROC curve for a system that operates at random. In other words, any verification attempt is given a 50% chance of acceptance, and a 50% change of rejection. Under these conditions, the false match rate equals the verification rate. Also indicated in Fig. 7.8 is a line where the false match rate equals the false non-match rate. This is not an ROC curve itself, but intersects ROC curves at their equal error rate (EER) point.

Reading ROC Curves

When reading ROC curves, it is common to first examine the x-axis, and select an acceptable rate of false matches. Next the verification rate is determined by moving up the graph vertically from that point. The reason is that for live systems, the matching threshold is usually selected to fix the false match rate (the probability of incorrectly matching an impostor) at the maximum acceptable level. In many, but not all, systems incorrectly refusing a genuine match is an inconvenience, but falsely matching an impostor can be a security risk. Therefore, the false match rate is fixed at a low rate such as 0.01%, and the verification rate is treated as a dependent variable.

7.2 Identification

The previous section dealt with the problem of verification, where users attempt to prove their identity to the system. In other words, the verification system answers the question "Am I who I claim to be?". In some situations, there will be no claim of identity. For example, assume a robbery has taken place, and CCTV surveillance footage has captured a clear picture of the perpetrator. In this case, the police will be interested in "Who is this person?". An *identification* system can be used to match the image against a database of known criminals in an effort to determine the identity of the suspect.

Other modes of operation for biometric systems are also possible. The preceding discussion has assumed that claims of identity are positive. However, negative claims of identity are also used for some applications. For example, consider a detention center where the authorities wish to know if a new inmate has been previously incarcerated, even if they provide a false identity. Assume a fingerprint image is captured for every new inmate. Before being added to the database, it is matched against all existing enrollments and likely matches are flagged. In this case, the implicit claim of identity is negative as people are claiming *not* to have been previously enrolled.

Depending on the type of biometric system being evaluated, different performance measures will be appropriate. In fact, in some circumstances it will not even be possible to compute standard verification error rates. For example, if an identification system only logs the top 20 ranked matches for an identification attempt, it will not be possible to compute verification error rates, as they rely on knowledge of the full genuine and impostor match score distributions (not just the scores of top ranked matches). This section discusses the performance measures appropriate for identification systems.

7.2.1 Identification vs Verification

Verifications involve a one-to-one match between a biometric sample and an enrollment template. On the other hand, identification involves a one-to-many match, with a biometric sample being matched against an enrollment database.[5] In fact, verification can be viewed as a special case of identification in which the system only has a single enrollment.

The terminology used for the samples and enrollments may vary between applications. A *probe* is a biometric sample from a user who's identity is unknown, and to be determined. The set of people enrolled in the system is known as the *database*, *gallery* or *watchlist*.

Identification systems often involve an operator who manually examines results. For example, consider an Automated Fingerprint Identification System (AFIS) that is used to identify latent fingerprints. In this case, the system is used to narrow down the entire database (which may contain millions of prints) to a small list of likely matches that can be manually verified. A manual examination of the result set is usually necessary, as the output of automated systems has no legal bearing (see Sect. 10.2). Furthermore, as with many identification scenarios, there is no guarantee that the probe even belongs to the gallery. This is known as open-set identification, and is the subject of Sect. 7.2.2.2.

The list of potential matches returned is known as the *candidate list*. There are two basic approaches to deciding which results to include in the candidate list. The system can either display all matches with a score above a given threshold, or it can display the X top ranked matches regardless of score. In either case, a system operator can adjust how many matches will be included on the list by either lowering the match threshold, or increasing X. The optimal number of matches to include in the list will depend heavily on the application. At one extreme, for many law enforcement applications (such as the AFIS example above) the operator will be willing to manually examine a large number of non-matches. However, they want

---

[5] Normally, the sample is matched against the entire enrollment database. However, template preselection schemes can be used to select a subset of the enrollment database for matching. See Sect. 7.4.2 for a discussion of error rates when the sample is not matched against the entire database.

to avoid a situation where true matches are not included on the candidate list, as they will not be seen. Therefore, they will set their thresholds relatively low. At the other extreme are high-security applications where false matches may be critically dangerous, while false non-matches are merely inconvenient.

7.2.1.1 Dependence on Number of Enrollments

In terms of performance, the key distinction between verification and identification system is the dependence of identification systems on the number of people enrolled in the system. For verification, a claim of identity ensures that the match will be one-to-one. Therefore, it does not matter how many people are actually enrolled. However, since identification systems conduct a one-to-many match, there is a direct relationship between the number of people enrolled and the probability of a false match. Intuitively, the more people who are enrolled, the greater the probability of there being a false match because of an enrollment with coincidental similarities to the input.

The dependence of identification performance on database size can be demonstrated more formally. Recall from Sect. 7.1.2.1 how score distributions underpin system performance. In general, the more overlap between the genuine and impostor distributions, the greater the probability of a matching error. The match score distributions for an identification system are now examined. Without loss of generality, assume that when a probe image is submitted for identification a candidate list is returned that consists of the single top ranked result (the same argument applies for the top X results). For the moment, assume that the probe person is not enrolled in the database. Now consider how the "Rank 1" impostor score histogram varies as the database size increases. When the database is of size 1, the problem is equivalent to verification as a one-to-one match is conducted. In this case, the Rank 1 impostor score distribution is the same as the original impostor score distribution (Fig. 7.9 (a)). Alternatively, if there are two people enrolled in the system, the probe will be matched against both, and the greater of the two scores will be returned. In this case, the Rank 1 impostor distribution of scores is expected to be a little higher, since the larger is always returned (Fig. 7.9 (b)). As the number of enrollments grows, the value of the top ranked score will continue to increase because there are more and more impostor scores to sample from (Fig. 7.9 (c)). These scores are all drawn from the same original impostor distribution, but since the maximum score is selected as the Rank 1 result, they are biased towards the high-end of the distribution. Now assume that the probe person is enrolled in the database with a single template. In this case, the probe will be matched against the corresponding template only once. Therefore, unlike the Rank 1 impostor score distribution, the Rank 1 genuine match score distribution does not change. Consequently, as the database size grows, the Rank 1 impostor distribution encroaches upon the genuine distribution (Fig. (d)), and the overlap between the two increases. The consequence of this is that as the top ranked impostor scores begin to trespass in genuine territory, false matches become more common.

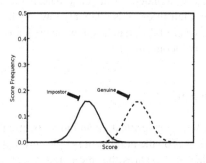

(a) The original genuine and impostor score distributions.

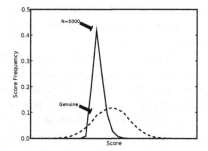

(b) The Rank 1 impostor score distributions for databases of size 1 and 2.

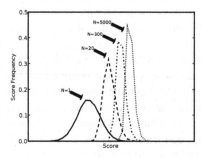

(c) Rank 1 impostor score distributions for a range of database sizes.

(d) The Rank 1 impostor score distribution for a database of size 5000, along with the genuine score distribution.

Fig. 7.9 The effect of database size on Rank 1 impostor distributions.

7.2.1.2 Score Normalization

Another difference between verification and identification is that score normalization can be used for identification. When the members of the gallery are known in advance, score normalization can be used to better distinguish between the subjects. For example, one approach is to apply a transformation to the feature space that maximizes inter-person distances and minimizes intra-person distances. In other words, the enrollments belonging to the same person are brought closer together, while those belonging to different people are distanced from each other. This will typically enable the algorithm to better discriminate between two people who were originally very close in the feature space. This approach works best when a) the gallery is static and relatively small (there are diminishing returns with large gallery sizes), b) there are no labeling errors in the database, and c) it is known that all people in the probe set are enrolled in the system (i.e. closed-set identification). In practice, these requirements are rarely met for systems in production.

Score Normalization and Testing

Score normalization is commonly used in commercial face and speech recognition identification systems. It is important for evaluators to know when it is in use because it ties match scores to a specific enrollment database. If scores are being normalized, it is generally not valid to combine match scores from different experiments. For example, one would not be able to generate the genuine score distribution using one dataset, the impostor distribution from a another dataset, and combine the results to determine system error rates. Furthermore, care must be taken when determining Rank 1 impostor distributions (e.g. for an alarm curve - see Sect. 7.2.2.2). It is important that only probes who do not exist in the gallery are used to determine these scores. In particular, one cannot use probes that exist in the database and simply select the highest impostor match score. This is because the existence of the probe in the gallery will have an influence on the impostor match scores (in particular, it will tend to drive them down). In general, when there is complete control over the testing protocol, it is recommended to enforce a policy that score normalization is not permitted.

7.2.2 Identification Performance

The differences between verification and identification systems were discussed in the previous section. On account of these differences, it is not surprising that the performance metrics used to evaluate and compare identification systems differ from those used for verification systems. With verification systems, the focus is on the outcome of a single match. However, for identification systems there are many implicit matches, any one of which can contribute to an identification error. There are two basic types of errors for identification systems:

False-negative identification error   When a user enrolled in the system does not have their true template returned in the candidate list after an identification attempt.

False-positive identification error   When a candidate list template is an impostor. In other words, each template returned by an identification query that does not belong to the user being identified is a false-positive error. For open-set identification, a false-positive error is often referred to as a *false alarm*.

The candidate list generally consists of all matches above a given threshold, or the top ranked X matches. Therefore, it is of interest how well an identification system places genuine matches in this list:

False-negative identification error rate The proportion of identification attempts in which a user enrolled in a system does not have at least one of their templates returned among the candidate list.

False-positive identification error rate The proportion of identification attempts in which a non-empty candidate list is returned, which does not include any genuine matches.

There is some room for interpretation in the definitions for these rates. For example, imagine a situation where a candidate list is returned with three results, of which the correct match is ranked third. On one hand, the identification is positive as the correct match belongs to the candidate list. However, incorrect matches are also included, and are ranked higher than the correct match. Is this both a correct identification and a false-positive? There is no definitive answer to questions such as this, and the resolution depends largely on the intended application. An alternative, and more specific definition, is that of a "Correct Detect and Identify". The requirement for this is that the genuine match is both a) in the candidate list, and b) ranked first. This issue is discussed further in the chapter on surveillance systems, in Sect. 11.4. Above all, it is important that an unambiguous definition is adopted before testing an identification system. This definition should be stated clearly and prominently in any report where performance rates are presented.

Reporting Identification Rates

Identification performance is closely related to database size. Therefore, when reporting an identification result, it is important to explicitly state the database size used for the experiment. For example, "The identification rate at a match threshold of 22 is 88% for a database of size 2000."

7.2.2.1 Verification-Based Performance Metrics

Verification performance statistics give probabilities for the two types of matching errors. Since identification systems are built upon a series of individual matches, it is possible to predict identification rates using verification rates, when available.

Assume a database of size N, where each user is enrolled with m templates. An input sample is matched against all enrolled templates, and the candidate list consists of all enrollments that achieved a match score above a system threshold t. Let FMR be the false match rate at t, and FNMR be the false non-match rate at t. The identification system false non-match rate is estimated as:

$$FNMR_N = FNMR^m$$

All m templates for the input person must falsely non-match in order for the identification to fail. Therefore, the more enrollments per person, the smaller the likelihood of a false non-match. In the case that there is a single enrollment per user, the identification false non-match rate is the same as for the verification case (because a single genuine match is conducted). Note that this assumes that all enrollments are matched. If the identification uses indexing, or the search aborts when the first match above t is found, the probability that the match is not conducted must be taken into account [5].

The probability of a false match is estimated as follows:

$$\text{FMR}_N = 1 - (1 - \text{FMR})^N$$

One or more false matches will cause an overall identification error. The probability of no false matches for a database of size N is $(1 - \text{FMR})^N$. Therefore, the probability of one or more false matches is $1 - (1 - \text{FMR})^N$. This is another demonstration that false match rates increase with database size.

It is important to note that these formulas only estimate error rates. They do not take into account intra-user correlation (i.e. varying error rates among users and user populations), and so they may be inaccurate for small databases. Therefore, whenever possible, identification error rates should be determined directly. This involves setting up experiments which involve full cross-matches between a set of probes, and a database of enrollments. The top genuine rank for each probe is examined to determine false non-match rates, and the top ranked impostor matches are examined to determine false match rates.

7.2.2.2 Open-set vs Closed-set Identification

There are two different types of identification systems: closed-set and open-set.

Closed-set identification Every input is known to have a corresponding enrollment in the database.
Open-set identification The sample submitted to the system may or may not have an enrollment in the database. Open-set identification is also known as the watchlist problem.

The distinction between the two is important, and impacts the way the corresponding systems operate, evaluations are conducted, and results are analyzed.

Recall that the candidate list is a list of potential matches returned as a result of an identification query. As mentioned previously, there are two ways to determine the candidate list: by using a match threshold, or by using a rank. For the first approach, only matches above the threshold are returned, and for the second approach, the top X ranked matches are always returned. The most suitable approach is dictated by whether or not the system is open-set or closed-set. In the case of closed-set identification, it is known that an enrollment exists, and the question is "how many of the top ranked matches need to be examined before finding the correct match?". On the other hand, for open-set identification it is not known if the enrollment exists in the database or not. Therefore, returning more matches does not necessarily increase the chance of a correct identification. In this case, a threshold based approach is more appropriate because one is only interested in likely matches. Ideally, if the sample does not have a corresponding enrollment, an empty candidate list is returned.

Rank-based and threshold-based analysis are conducted in different ways. However, they are both impacted by the dependence on the number of enrollments (see Sect. 7.2.1.1). In the case of ranks, consider that the top X ranked scores are returned. In this case, at the database size grows, the number high-scoring impostors increases, and these "push" the genuine matches out of the candidate list. On the other hand, consider candidate lists that are defined by a threshold. Once again, as the database grows, there will be a greater likelihood of high-scoring impostor matches. However, this does not change the likelihood that a genuine match is included in the candidate list, as it will always be included if it scores above the threshold. However, it does mean that non-empty candidate lists will become more common for people who are not enrolled in the system. In other words, the rate of false alarms will increase.

The primary graph for showing rank-based performance is the CMC curve, and the primary graph for threshold-based performance is the alarm curve.

Closed-set Systems and the Real World

Closed-set systems generally only exist in laboratory situations, such as technology evaluations. In general, making the assumption that every input belongs to someone enrolled in the database is unfounded (and can even be dangerous) because in real-world situations it is difficult to enforce this policy. There almost always exists some possibility (e.g. through human error or a security breach) that the input belongs to someone unexpected and unauthorized. However, the results from closed-set evaluations are still of interest as they do reflect system identification capabilities, and can expose system weaknesses.

7.2.2.3 The CMC Curve

The most common graph for evaluating closed-set systems is the *cumulative match characteristic* (CMC) curve, which depicts identification rates over candidate list sizes. In this case, the candidate list is selected by including the top X ranked results. This is a typical approach for closed-set identification, because it is known that the input sample has a corresponding match in the database - the performance metric of interest is where the genuine match is ranked. Ideally the rank is high, so fewer matches must be examined to locate it.[6]

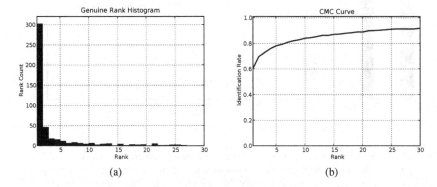

Fig. 7.10 Genuine rank results for a system with 500 identification transactions, and a database of 5000 enrollments. (a) A histogram of genuine ranks. For about 300 of the identifications, the genuine match ranked at position 1. Approximately 50 were ranked at position 2, and so on. (b) The corresponding CMC curve. At each rank, this is the proportion of genuine matches that ranked at or below this point. For example, around 60% of genuine matches were rank 1, and almost 80% were ranked in the top 5.

In order to compute a CMC curve, the first step is to tabulate the ranks of all genuine matches. For example, observe Fig. 7.10 (a). This shows the distribution of genuine ranks scores for a system with 500 probes, and a gallery of size 5000. For about 300 identifications, the correct match was ranked first. Therefore, the *rank 1 identification rate* is approximately $300/500 = 60\%$. This is the first point plotted on the CMC curve in Fig. 7.10 (b). The rank 2 identification rate is the proportion of genuine matches that ranked either 1 or 2. In Fig. 7.10 (a) it can be seen that there are about 50 matches with rank 2. Therefore, the rank 2 identification rate is approximately $(300 + 50)/500 = 70\%$, which is also plotted in the CMC curve. In general, the rank r identification rate is the proportion of inputs whose enrollment is ranked within the top r matches. Since closed-set identification is assumed, the identification rate at rank N (where N is the database size) is guaranteed to be 100%. In other words, if the whole database is returned, the correct match is always in the candidate list.

Since CMC curves are cumulative, they are monotonically non-decreasing (i.e. it can stay constant or increase, but cannot decrease). Furthermore, since the rank N rate is 100%, the curve converges towards 100% as the rank increases. For an accurate system, CMC curves start high and converge to 100% quickly.

---

[6] Ranking terminology can be confusing, as a the rank 1 result is considered "high", while the rank 500 is considered "low". There is an inverse relationship between the rank number and the match score.

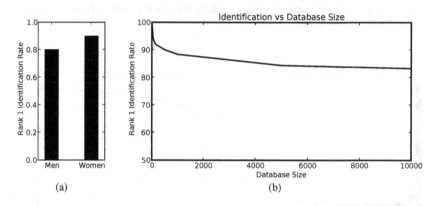

Fig. 7.11 Rank 1 identification graphs. (a) An example of a graph that compares rank 1 identification rates for men and women. (b) The decrease in rank 1 identification rates as the database size increases. For example, for a database of size 700, the rank 1 identification rate is roughly 90%. However, the same algorithm has a rank 1 identification rate of around 85% for a database of size 4000.

Another graph that can be useful for the evaluation of identification systems is the Rank 1 (or Rank X) graph. These graphs plot identification rates across different conditions. For example, they can be used to compare identification performance between demographics, such as men and women (see Fig. 7.11 (a)). Also, they can be used to depict how identification rates vary as database sizes increase, as in Fig. 7.11 (b). As database size increases, the chance that the genuine enrollment is ranked first decreases. Note that the performance decrease is not linearly related to the increase in enrollments, but rather log-linear [18].[7]

7.2.2.4 The Alarm Graph

With closed-set identification, a correct match is guaranteed as long as a sufficient number of the top ranked matches are included in the candidate list. Therefore, for every input, a non-empty candidate list should be returned. However, for open-set identification, it is not known if the input sample has a corresponding enrollment or not. Therefore, an empty candidate list is acceptable, and indicates that the input person is unlikely to be enrolled. In this situation, a match threshold is used to determine candidate list membership, rather than ranks.

There is specific terminology that is often used for open-set identification. An *alarm* is sometimes used to describe a non-empty candidate list for open-set identification based on a match threshold.[8] The list of people enrolled in the system is the

---

[7] For a log linear scale, if the x-axis is plotted logarithmically, the relationship appears linear.

[8] The term *alarm* is sometimes used to describe the top ranking result, regardless of membership in the candidate list.

watchlist. When there is a operator for a live identification system, an alarm alerts them (often audibly) that a likely match has been found. The match score threshold is known as the *alarm threshold*. When an alarm is raised for an individual enrolled in the system it is known as a *detection*, and when they are not enrolled in the system, it is known as a *false alarm*. An *identification* occurs when the correct match is ranked number 1, regardless of score. A *correct detect and identify* is when the genuine match is both ranked 1, and above the alarm threshold. When evaluating an open-set system, one must determine what the desired outcome of a query is: a detection, or a correct detect and identify. Generally speaking, a correct detect and identify is a "better" outcome, because with a detection there is a possibility that the true match is not ranked first. However, if one assumes that an operator is willing to examine all alarms (not just the top ranks), they will be presented with the correct match. Therefore, in most cases a detection is sufficient. Furthermore, the analysis is easier if it is focused on detections.

There are two performance rates of interest. The first is the probability that the correct match, when it exists, is included as part of the candidate list. This is known as the *detection rate*. Since inclusion in the candidate list is based only on the alarm threshold, this rate does not depend on database size. If a genuine match scores above the threshold it is included in the candidate list, regardless of ranking and how many impostor matches are also above the threshold. The second performance rate is the *false alarm rate*, which is the proportion of identification queries that raise false alarms.

What Constitutes a False Alarm?

If both genuine matches and impostor matches are included as part of the candidate list, is this still a false alarm? The authors recommend the following resolution. Detection analysis and false alarm analysis should be conducted separately (this is discussed further in Sect. 11.4.3). Detection rates are computed based on probes who belong to the watchlist, and false alarm rates are based only on probes who do not have corresponding enrollments in the database. The reasons for this is due to score normalization (see Sect. 7.2.1.2). The presence of a genuine match in the database can actually alter the score distribution of impostor matches. Therefore, an event is only considered a true false alarm for non-watchlist individuals, and hence a mixed candidate list is not considered a false alarm.

The false alarm rate depends on database size. Basically, as more people are enrolled, the more impostor matches that are conducted, and the greater the probability that there will be at least one above the threshold and included in the candidate list.

An *alarm graph* or *watchlist ROC curve* plots the detection rate against the false alarm rate over a range of alarm thresholds. It is computed in a similar manner

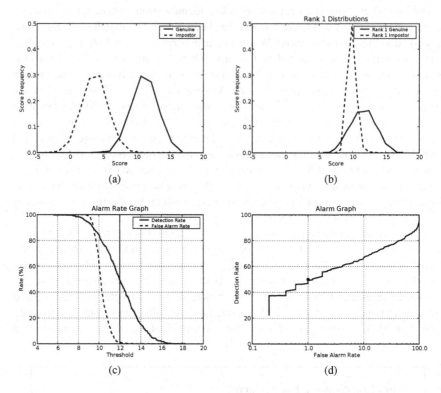

Fig. 7.12 The basis of the alarm curve. (a) The genuine and impostor match score distributions for a hypothetical algorithm. (b) The rank 1 genuine distribution and the rank 1 impostor distribution. These distributions are used because it is the highest score that determines if the match raises an alarm. These distribution are related to the number of enrollments. (c) The detection rate and false alarm rate. For example, at threshold value of 12 (indicated by a vertical line), the false alarm rate is about 1%, and the detection rate is about 50%. (d) The corresponding alarm graph. Each point represents a specific alarm threshold. For example, the point at a false alarm rate of 1% and identification rate of 50% (indicated by a dot) corresponds to a threshold of 12.

as the ROC curve.[9] However, instead of being based directly on the genuine and impostor distributions, it is based on the rank 1genuine distribution and the rank 1 impostor distribution (see Fig. 7.12 (b)). The reason that it is based on the rank 1 distributions is that only the highest ranking score is of interest, because as long as there is a single genuine/impostor score above the threshold, this is sufficient for a detection/false alarm. Note that this is what gives alarm rates its dependence on database size. As the watchlist grows, the rank 1 distributions will move to the right (see Fig. 7.9). The detection rate is computed as the proportion of rank 1 genuine

---

[9] The key difference between an ROC curve and an alarm graph is that the latter is for identification systems, and is therefore dependent on database size. In fact, the ROC graph is identical to an alarm graph when the enrollment database has size 1. In this case, the probability of a false alarm is the same as the probability of a false match, and identification rates are also equivalent.

scores above a threshold, and the false alarm rate is the proportion of rank 1 impostor scores above a threshold. Both of these rates drop as the alarm threshold increases, as illustrated in Fig. 7.12 (c). The alarm graph, Fig. 7.12 (d), plots the detection rate/false alarm rate pairs over a range of thresholds. In general, points toward the upper-left of the graph represent high performance, as this is where the false alarm rates are low, and the detection rates are high.

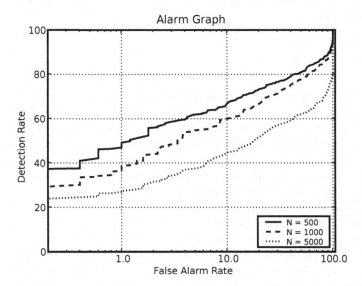

Fig. 7.13 Sample alarm curves for a recognition algorithm applied to different sized database sizes. At a false alarm rate of 1%, the detection rates are approximately 37%, 40% and 50% for the same algorithm as applied to databases of size 500, 1000 and 5000 respectively.

Figure 7.13 contains a series of alarm curves for the same algorithm, as applied to different database sizes. This illustrates how the alarm curve is a function of watchlist size. Note how the identification problem becomes more difficult (i.e. the detection rates drop) as the database size increases. This is illustrated by the lower curves for database sizes of 1000 and 5000. In fact, it appears the curves are being shifted down and to the right.

7.2.3 Projecting Results to Large Database Sizes *

It is often impractical to run experiments with databases as large as would be expected for a live system. Consider a trial designed to estimate identification rates for a database that is expected to contain in excess of one million enrollments. For example, the database may contain all driver's license photos for a country, with thousands of new images added daily. In this case, it is not feasible to recruit millions of test subjects for a full scale test. However, without an estimate of expected identification rates as the database grows, the driver's license authority would be hesitant to invest in the large-scale project. Therefore, projecting results is an important area of research. However, the best method for projecting results to arbitrary database sizes is still an open problem, and is being actively discussed in the literature. An in-depth analysis of all the issues is beyond the scope of this text, however an overview is presented. References [2, 9, 10, 12, 18] are recommended to the interested reader.

7.2.3.1 CMC Curves

In order to compute true CMC curves, it is necessary to know the rank r identification rate for $r = [1,...,N]$. Therefore, in order to predict a CMC curve for arbitrary database sizes, it is necessary to estimate the rank r probabilities as a function of database size. The formulas in Sect. 7.2.2.1 compute the overall false match and false non-match rates for a database of size N for a threshold t, but do not give rank-based identification rates. For example, they cannot be used to determine the rank 10 identification rate for a database of size N. This is more complicated, as it involves both false match and false non-match information.

Consider a closed-set identification system with N enrollments, with exactly one enrollment per user. Since it is closed-set, there must be a corresponding match in the database for any input. Assume a genuine match score s has been achieved for a particular input. We are interested in the following: what is the probability of a correct rank 1 identification? In order for this to occur, the score s must be greater than all N-1 impostor match scores. Let $g(s)$ be the genuine score probability density function and $I(s)$ be the cumulative distribution function for impostor scores (i.e. the probability of an impostor score less than s). The probability of s being rank 1 is the probability of having N-1 impostor scores less than s:

$$\text{Rank1}(N,s) = I(s)^{N-1}$$

Now consider an arbitrary s. To find the overall rank 1 probability, integrate over all possible genuine scores:

$$\text{Rank1}(N) = \int_{-\infty}^{\infty} g(s)I(s)^{N-1}ds$$

This formula allows us to predict the rank 1 identification rate based on the genuine and impostor score distributions, which can be determined from a relatively small data set (thousands of samples as opposed to millions of samples). Next, the formula is generalized to other ranks. At rank r, there are r-1 impostor matches with scores above s, and N-r impostor matches with scores below s. The Bernoulli coefficient gives the number of possible combinations of the match scores for this to happen, so the probability of identification at exactly rank r is:

$$p(N,r) = \int_{-\infty}^{\infty} \binom{N-1}{r-1} g(s) I(s)^{N-r} (1 - I(s))^{r-1} ds$$

For identification at rank r, the genuine match can be ranked anywhere from 1 to r. Therefore, the rank r identification rate is:

$$\text{IdentificationRate}(r) = \sum_{x=1}^{r} p(N,x)$$

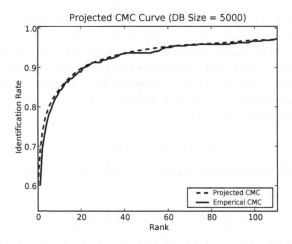

Fig. 7.14 An empirical CMC curve and a projected CMC curve. The curves have been generated using synthetic data, and therefore all underlying match scores are independent and identically distributed.

In theory, the final formula gives the identification rate at any rank. Therefore, it can be used to generate a CMC curve. Figure 7.14 contains two CMC curves for synthetic data. One curve is generated empirically, by tabulating the ranks of the genuine match scores and computing the identification rates directly. The other curve is generated using the projection method. As can be seen, the two curves match very closely.

The method outlined above has known limitations. One key problem arises from the assumption that all impostor scores are independent, and that all users have identical genuine match score distributions (this is the case for the synthetic data for Fig. 7.14). It is well known that these assumptions are false as individual users tend to have their own match score distributions (see Chap. 8). This information is not captured by system-wide match score distributions, which are used as the basis of the projection. The result is that the predicted identification rates consistently underestimate the empirical identification rate. Several studies have demonstrated that normalizing user score distributions before applying the projection improves the result [9, 18]. Furthermore, there are numerical computation issues that require the use of specialized algorithms to avoid precision errors and reduce the running time.

7.2.3.2 Alarm Graphs

Assume that genuine and impostor score distributions are independent and identically distributed for all users. If they are known with enough precision, one has complete knowledge of a biometric system. Therefore, one should be able to predict performance under a variety of different conditions, including identification scenarios.[10]

The core feature of identification systems is that they match an input against a database with N enrollments. The formulas in Sect. 7.2.2.1 have a parameter N. Therefore, they can be used to give estimates for the false match rate and false-non match rate at a given threshold for arbitrary database sizes. Therefore, these can be used to create alarm graphs for large database sizes. Assume a database of size N, and people are enrolled with a single template. At an alarm threshold of t, the true-positive identification rate (IR) is:

$$\mathrm{IR}_N(t) = 1 - \mathrm{FNMR}(t)$$

and the false-positive identification error rate (FPIR) (also known as the false alarm rate) is:

$$\mathrm{FPIR}_N(t) = 1 - (1 - \mathrm{FMR}(t))^N$$

For the projected alarm curve, false alarm rate is plotted against the identification rate for a range of thresholds. At a database of size 1, this graph is equivalent to the ROC curve. In fact, the effect of larger databases is that the ROC curve is shifted towards higher false alarm rates.

7.3 Dealing with Uncertainty

No matter how carefully one conducts an analysis, the error rates established through testing will always differ from the observed error rates for the same system in operation. Uncertainty is an important topic for biometrics, and has important implications for designing test plans and reporting results. The ultimate goal is for all results to be "statistically significant", meaning they are unlikely to be due to chance.

There are two primary sources of errors that prevent one from knowing "true" performance statistics: systematic errors and sampling errors. There is little that can

---

[10] This statement makes an implicit assumption that all factors apart from database size are held constant. For example, if population demographics or enrollment quality change, this will affect performance in ways that are difficult to predict.

be done to quantify systematic errors, but the magnitude of the sampling error can often be estimated. Confidence intervals are used to present the range of uncertainty. A related problem is estimating the size of the test population needed in order to obtain results of sufficient precision.

7.3.1 Systematic Errors

Systematic errors result from environmental factors or test procedures that bias the result in any direction. An example of a systematic error for a biometric system is when the demographics of the test subject population does not match that of the target population (e.g. young males were over-represented because they were more likely to respond to the advertisement for participants). Ideally, test subjects should be randomly selected from the target population. However, this is not always possible. Behavioral aspects can also introduce systematic errors. For example, if a test subject for a covert surveillance system knows the nature of the trial, he or she may alter their behavior (e.g. by looking for the hidden cameras). In theory, systematic errors can be identified and controlled; in practice they are inevitable. The difficulty lies in the fact that they are always present and bias the result in the the same direction, so they cannot be uncovered or removed by simply running the experiment multiple times. Therefore, more often that not, they remain unknown and unobserved. The best approach for dealing with systematic errors is through careful experiment design, and strict adherence to best practice standards [11, 15].

7.3.2 Sampling Errors

Sampling errors are a consequence of using a finite set of subjects to represent a larger target population. Sampling errors are always present, even when the subjects are selected randomly. As will be seen in Chap. 8, unique individuals tend to perform differently within a biometric system. For example, some people may have trouble authenticating, while others may be prone to being falsely accepted. Therefore, the exact set of people selected to participate in a trial will have some influence on the collective error rates. Unlike systematic errors, statistical theory can be used to estimate the degree of sampling error. However, this assumes that the trial participants have been selected randomly, and are representative of the target population. Sampling error is inversely related to sample sizes; the more people involved in a trial, the smaller the sampling error.

A crucial consequence of sampling error is that all reported performance measures have an inherent degree of uncertainty. The size of this error must be taken into consideration when comparing results from two or more systems. A result is called *statistically significant* if it is unlikely to be due to sampling error.

Gaining Confidence in a Result

The following is a simple example to illustrate the concepts of sampling error and confidence intervals. Consider two iris verification systems A and B, which are under evaluation. The goal of the evaluation is to determine which system has a higher verification rate. Each system is tested by having 5 people attempt a verification, and recording the result. Assume that system A correctly accepts 3 of the 5 people, and system B correctly accepts 4 of the 5 people. A naive interpretation is that system B's verification rate is 20% higher than that of A. To verify this, the experiment is run a second time with a different 5 people. This time system A has a 100% verification rate, while system B's error rate drops to 60%. With this additional information, it is not clear which system is better. With such a small sample size, it is not surprising that the results vary so widely. However, as the test is repeated more times with a greater number of people, trends will become apparent. Gradually, *confidence* is developed that the true performance rate falls within a given range. This range is given as a confidence interval. The greater the sample size, the smaller the range. If enough samples are used, the confidence interval for system A will not overlap with the confidence interval for B. If these two ranges do not overlap, one can be pretty sure which is the better system.

7.3.3 Confidence Interval Interpretation

A confidence interval is an interval estimate $[L, U]$ of an observed error rate, along with an associated probability p that the true error rate is between L and U.[11] For example, instead of only reporting that the observed false match rate was 0.01%, one may also provide a 95% confidence interval [0.001%, 0.05%]. The reason for the confidence interval is that the true error rate is unlikely to be exactly 0.01% due to sampling error. However, it is likely to be *close* to that value. Confidence intervals formalize this concept.

Confidence intervals are especially useful when comparing results. When sample sizes are small, there is little certainty about the true value of the error rate. Consequently, care must be taken when comparing two results as the absolute distance between the measurements may be misleading. The following heuristic can be used

---

[11] Technical note: Strictly speaking, this is not the correct interpretation of a confidence interval. Probabilities cannot be inferred about the true error rate as it is not a random variable, and therefore does not have a probability distribution. The correct interpretation is that if the experiment is repeated many times, p is the proportion of confidence intervals that would contain the true parameter. The difference is subtle, and has generated much unnecessary confusion.

for comparison: if confidence intervals of the two measurements do not overlap, there is a statistically significant difference between the values. Of course, the size of the confidence intervals is important. The statement holds true for 95% confidence intervals, but this is actually overly conservative [17]. Confidence intervals of at least 85% should be used in practice.[12]

Figure 7.15 contains two graphs, each containing two ROC curves with 85% confidence intervals indicated. In the case of Fig. 7.15 (a), the confidence intervals are narrow, implying that the ROC curves are very reliable. The intervals between the two curves do not overlap, so the claim that "System A has higher verification rates than System B" is statistically significant and thus unlikely be to an artifact of sampling error. In contrast, in Fig. 7.15 (b), the confidence intervals are relatively wide, and there is overlap between the two curves. The verification rates for System A are higher than System B. However, since the confidence intervals overlap, the result may be a result of sampling error, and another test may have had a different outcome. Therefore, one cannot make a statistically significant claim that System A is better than System B without further testing.

7.3.4 Computing Confidence Intervals *

Computing confidence intervals for biometric error rates is an important, but complicated subject. To a large extent, the most appropriate method for computing confidence intervals for biometric data remains an open problem and is an active area of research. The following is a discussion of some of the issues involved, however it is not an exhaustive survey of the field. The reader is encourage to follow the references provided where a deeper understanding is required.

Consider a biometric system that is being evaluated to determine its false rejection rate. The test population consists of N people, each of whom attempt a single verification. For each verification attempt there are two possible outcomes: they may be correctly accepted, or falsely rejected. Let p be the proportion of false rejections. At first glance, this appears to be a *binomial experiment*, which has the following requirements:

1. There is a fixed number of trials
2. There are two possible outcomes for each trial (accept or reject)
3. Each trial is independent
4. The outcome probabilities are constant

Interval estimation for binomial proportions has been studied extensively in statistics. The problem is not trivial, as there are complications arising from the discrete

---

[12] Non-overlapping 83% or 84% confidence intervals give an approximate $\alpha = 0.05$ test, therefore one should use at least 85% confidence intervals or higher. However, this makes some assumptions about the standard errors of the test statistics, so should be considered a rough heuristic rather than a formal rule.

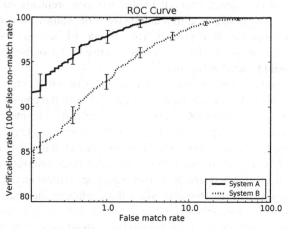

(a) Non-overlapping confidence intervals.

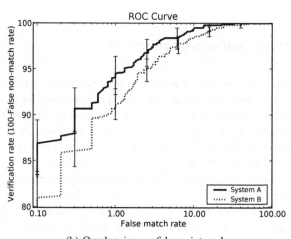

(b) Overlapping confidence intervals.

Fig. 7.15 ROC curves with 85% confidence intervals. For the two curves in (a), it is clear that System A has a statistically significant improvement in verification rates as there is no overlap of the confidence intervals. For the two curves in (b), System A appears to have better performance, but further testing is necessary to make the result statistically significant.

nature of the binomial distribution that cause the coverage of some standard approaches to behave erratically [4]. A simple approach involves approximating the binomial distribution by a normal distribution (with mean p and variance $p(1-p)/N$), as justified by the central limit theorem. However, this approach is not appropriate for biometric data, as it is inaccurate when the p is too close to 0, and biometric error rates are sometimes very small (e.g. $p \ll 0.001$). Furthermore, assuming normality results in a symmetric interval. However, when the proportion is near 0 this may result in negative probabilities, which are not possible. Therefore, the use of normal approximations should be avoided. An alternative approach is to use an exact method, such as the Clopper-Pearson interval, which are based directly on the binomial distribution [6]. However, these methods tend to be overly conservative, in the sense that coverage may be larger than necessary [1].

There are further, more fundamental, issues that must be considered. The conditions for a binomial experiment must be re-examined. Conditions 1 and 2 are fine, as there is a fixed number of trials with two well defined outcomes. Condition 3 is satisfied as long as each person only attempts verification once. However, this is often not the case. When conducting a trial, recruiting subjects is time consuming and expensive. Therefore, it is usually desirable to have each person attempt several verifications. However, these verifications will not necessarily be independent of each other as the subject's behavior may alter with successive attempts. Condition 4 is the most problematic. It assumes that the probability of success for each test subject is constant. However, it is well known that in many biometric systems individual error rates vary (this is the subject of Chap. 8). In other words, the scores for a given user are correlated, and this is sometimes referred to as as intra-user correlation. The application of interval estimation for binomial proportions based on identical genuine and impostor match score distributions for the whole population will give misleading results. In general, the variance (uncertainty) of the statistic will be underestimated, resulting in a confidence interval that is too narrow. In other words, it will appear that there is more confidence in the results than there actually is.

There are two requirements for a method of confidence interval estimation: it must allow for a varying number of attempts per subject, and it must take into account intra-user correlation. There are two main approaches: parametric and non-parametric.

7.3.4.1 Parametric Techniques

Parametric approaches are based on models, and make assumptions about the nature of underlying distributions. An example is the binomial proportion method discussed above. Another prominent method is to use the family of beta-binomials to model the correlation [22]. An advantage of model based approaches is that they can be very fast to compute. However, a disadvantage is that they sometimes rely on parameters that need to be tuned using large amounts of data. Furthermore, the assumptions of the model may turn out to be unfounded, as in the case of the binomial distribution.

7.3.4.2 Non-parametric Bootstrapping

The most commonly used non-parametric method for biometric error confidence intervals is a re-sampling method known as bootstrapping [3]. In bootstrapping, the original set of match scores is sampled with replacement to create bootstrap samples, and the parameter of interest is computed for each bootstrap sample. This distribution of the parameters can be used to determine confidence intervals. For example, consider the case of computing the false non-match rate for a trial with n subjects, each of whom has made m verification attempts. A bootstrap sample is created by the following process:

1. Sample with replacement n subjects from the test subject population.
2. For each subject, sample with replacement m scores. This gives us a total of $m \times n$ scores, which is different from the original result set.
3. Compute the false non-match rate for the bootstrap sample.

This is repeated many times (e.g. 1000), giving an empirical distribution for the false non-match rate estimator. To create a $100(1 - \alpha)\%$ confidence interval, the results are sorted, and the bottom and top $100 \times \alpha/2\%$ bootstrap values are discarded. For example, assume there are 1000 bootstrap samples, and a 95% confidence interval is to be computed. In this case $\alpha = 0.05$, so the top and bottom 2.5% (or 25 samples) are discarded. The remaining values give the range of the confidence interval.

The advantages of bootstrapping are that it is simple and intuitive, and few assumptions are made about the nature of the underlying data. The primary disadvantage is that it is computationally expensive. The bootstrapping process itself is time consuming, and large numbers of bootstrap samples are needed. Therefore, it may be infeasible for large test sets.

7.3.5 Boxplots

Another method for illustrating statistical uncertainty is by using boxplots, as in the 2006 FRVT competition [19]. Boxplots are similar to confidence intervals, except they contain more information as they give an indication of the underlying spread of results, as opposed to simply an upper and lower bound. For the FRVT competition, the boxplots are generated using a sampling technique. The test set is partitioned into smaller sets, and the performance rates are computed for each subset. For example, the test set may be divided into 30 smaller sets, and the false reject rate (FRR) at a fixed false accept rate (FAR) is computed for each subset. Boxplots are a non-parametric way to depict the resulting performance rates using five numbers to summarize the data: the smallest observation, the lower quartile, the median, the upper quartile, and the largest observation. Outliers are also included, as illustrated in Fig. 7.16 (a). By examining a boxplot one gets an impression of variance and skewness of the distribution. The area between the lower and upper quartile contains 50% of the results, so this region gives a region where the true result is likely to lie.

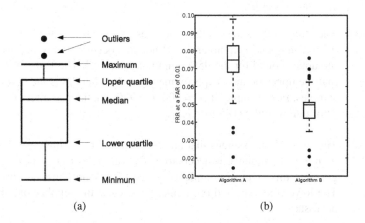

(a)   (b)

Fig. 7.16 (a) A labeled boxplot. (b) An example of two boxplots for hypothetical algorithms. 50% of false reject rates measured for Algorithm B fall in the range [4.5%, 5.2%] at a false accept rate of 1%. This is considerably lower than for Algorithm A, whose results fell in the range [6.9%, 8.3%] 50% of the time.

Boxplots can be used for visually comparing systems. Figure 7.16 (b) contains sample boxplots for two algorithms. In this case, one can be reasonably confident that Algorithm B has lower false reject rates, as the body of the boxplot is considerably lower than the body for Algorithm A.

7.3.6 Calculating Sample Sizes

It has been established that the size of a confidence interval is inversely related to sample size. In other words, the greater the number of people involved in a trial, the more reliable the error rate estimates. Therefore, a natural questions is: "How big should my sample size be in order to achieve my desired level of confidence?". To answer this question, one must first decide what level of confidence is desired. Furthermore, the number of samples required to reach this confidence level depends on the error rate itself. Therefore, it is also necessary to have an estimate of the error rate. In practice, it is difficult to have an accurate estimate of the error rate a priori. In fact, if it is known with precision, a trial to determine it would not be necessary.

For the discussion of sample sizes, it is assumed that each person authenticates only once. In general, having 10 people attempt 10 authentications is not equivalent to having 100 people authenticate once. This is because subsequent transactions from the same person are not statistically independent, so do not contribute as much new information as a test from an unseen user. In general, it doesn't hurt to have people authenticate multiple times, but sample size estimations should be based on the number unique users, not transactions.

There are two commonly used heuristics for sample size estimation: the Rule of 3 [13], and the Rule of 30 [20].

Rule of 3   The Rule of 3 addresses the worst case scenario of a trial being conducted and observing no errors. When this occurs, we are interested in the upper bound for the 95% confidence interval. The Rule of 3 states that the upper bound is approximately $3/N$, where N is the number of samples. For example, if $N = 300$, the 95% confidence interval for the error rate would be 0% to 1%.

Rule of 30   The Rule of 30 assumes that an error rate $\pm 30\%$ of the observed rate with 90% confidence is acceptable. The rule is as follows: in order to generate this level of confidence at least 30 errors must be observed. The lower the expected error rate, the greater the number of samples necessary.

It should be noted that these heuristics are only to be used as rough estimates. As they do not take intra-user correlation into account, they have been shown to underestimate the number of samples actually needed [7]. There are other methods in the literature for computing sample size requirements. For example, model based approaches for computing confidence intervals can sometimes be inverted to obtain necessary sample sizes [21]. However, beyond the mantra "the more the better", there is currently no universally accepted approach to establishing sample sizes. This continues to be an important area of research.

Rule of 3 and Rule of 30 Examples

Rule of 3: Assume a vendor would like to make a statistically significant claim that their algorithm has a false match rate less that 0.01%. In this case, the number of necessary independent impostor samples is 30,000 ($3/30000 = 0.0001 = 0.01\%$).

Rule of 30: If the error rate is expected to be around 1%, there should be at least 3,000 attempts, since $0.01 \times 3,000 = 30$. In this case, the 90% confidence interval would be $[0.7\%, 1.3\%]$. For an expected error rate of 0.1%, at least 30,000 attempts are needed.

7.4 Other Performance Measures

Sections 7.1 and 7.2 presented the standard error rates for verification and identification systems. All of these performance rates are built upon the two fundamental types of match errors: false matches and false non-matches. However, there are other types of errors that can occur within a biometric system. Enrollment errors and binning errors are discussed below.

7.4.1 Enrollment and Acquisition Errors

There are a number of reasons why a biometric system may fail to obtain a biometric sample. The exact reason why the biometric sample could not be obtained depends on the biometric being used, as well as the environment and hardware. For example, a dirty or greasy finger may lead to a failed acquisition for a fingerprint verification system. For face recognition, glare from eye glasses may make it difficult to locate the eyes, thereby preventing feature extraction.

In general, there are several necessary steps before matching can take place:

1. **Data capture:** The biometric is presented to a sensing device, which captures the information and converts it to an electronic format (e.g. a digital camera).
2. **Signal processing:** The biometric sample is processed to extract the distinctive features. Typical sub-steps include:

   a. *Segmentation* partitions the input into two regions: the foreground that contains the biometric sample, and the irrelevant background region. The foreground is maintained, and the background is discarded.
   b. *Feature extraction* scans the foreground for distinctive features that may be useful in discriminating between different biometric samples. This has the effect of reducing the dimensionality of the input, as samples may then be compared on the basis of these features instead of the entire input.
   c. The features are used to create a *template*, which is a compact representation of the features that will be used for matching.

3. **Storage:** During enrollment, the template is stored in the enrollment database.

Errors can occur at any of the stages outlined above, preventing a match from taking place. Depending on when the error happens, there are two types of errors:

Failure to enroll If the failure occurs during enrollment, it is known as a failure to enroll. The proportion of enrollment transactions that fail is known as the failure to enroll rate (FTE).

Failure to acquire If an error occurs while acquiring the biometric sample during a verification or identification attempt, it is known as a failure to acquire. The proportion of verification or identification attempts that fail for this reason is the failure to acquire rate (FTA).

For either type of failure, a comparison between the sample and enrollment template is not possible (because one or both are missing). Therefore, there is no evidence that the match is genuine, so the verification decision must be a rejection. This result must be distinguished from a false non-match since no similarity score has been generated. In other words, it is a system error, but not an algorithmic failure. System policy is usually designed to mitigate failures to enroll and failures to acquire events. For example, a user may be allowed to present their biometric three times during enrollment. If all of these attempts are unsuccessful, the result is a failure to enroll. The protocol that is defined for enrollment and verification or identification is known as a system's *decision policy*.

The errors discussed in the previous two sections have been concerned strictly with match errors. At a higher level, one can define errors at the transaction level that include both matching errors, and enrollment or acquisition errors.

False Accept A false accept occurs when a verification transaction confirms the
      identity of an impostor. In order for this to occur, a prior enrollment
      must have been successful, a sample is successfully acquired, and there
      is a false match between the acquisition sample and the enrollment tem-
      plate.
False Reject A false reject is a genuine verification that is incorrectly denied. This
      may be due to any of the following: a failure to enroll, a failure to
      acquire, or a false non-match.

A false accept is analogous to a false match, the difference being that a false accept occurs at the transaction level, and may be comprised of several matches. Similarly, a false reject is the transaction level equivalent of the false non-match.[13]

The following are the transaction level performance rates:

False Accept Rate (FAR) The proportion of system transactions that are falsely ac-
      cepted. In other words, this is the false match rate for samples that are
      successfully acquired.
False Reject Rate (FRR) The proportion of system transactions that are falsely re-
      jected. This includes failures to acquire, and the proportion of false
      non-matches for successful acquisitions.

These rates are computed as follows:

$$FAR = FMR \times (1 - FTA)$$

$$FRR = FTA + FNMR \times (1 - FTA)$$

The performance measure and graphs of Sect. 7.1 can be amended to use the false accept and false reject rates. For example, instead of plotting the false match

---

[13] It is not uncommon for the terms false match/false accept and false non-match/false reject to be used interchangeably. However, the distinction is important, even if it seems pedantic.

rate vs. the false non-match rate, an ROC curve can plot the false accept rate vs. the false reject rate.

Reporting Error Rates

A system can artificially increase its apparent performance for some error measures by rejecting a large proportion of the low-quality input. Therefore, when evaluating or comparing systems, it is vital to have a strict policy about failures to enroll and failures to acquire. All failures must be reported, and it is advisable to use performance measures that take into account enrollment and acquisition errors.

7.4.2 Template Pre-selection or Matching Subsets

Section 7.2.1.1 illustrated the dependence of identification performance on database size. In essence, the larger the database, the more difficult the identification task. This result assumes a full 1:N match against all enrollments. One approach to dealing with large databases is to only match against a subset of enrollments, thereby reducing the probability of a false match.

There are various approaches to selecting the database subset to match against. The exact method relies on the biometric being used. For fingerprint identification, a natural approach is to use fingerprint classification. There are 5 basic categories of fingerprints, based on the overall pattern of ridges (arch, tented arch, whorl, left loop, and right loop). Consider an input fingerprint that is a left loop. If the class of all fingerprints in a database is labeled, it makes sense to only match the input print against left loop enrollments. This has both speed advantages, as well as avoiding potential false matches. However, the classification task of fingerprints is very difficult. Some fingerprints have properties of two or more classes, making classification somewhat subjective. Therefore, there is a risk that if an input print is only matched against other prints of the same class, a classification error will cause it to be matched against a subset that does not contain the true match.

Another approach is based on using metadata to filter the set of enrollments. For example, a face recognition system may only match male inputs against male enrollments. In theory, this should not reduce the change of a correct verification, as the labels are usually well defined and mutually exclusive. However, large databases will always struggle with data integrity, and some level of labeling errors are inevitable. Therefore, in reality this may result in subsets being returned that do not contain the correct match.

The method employed for selecting which enrollment templates to match the input against is known as the *pre-selection algorithm*. The process is also known as classification, indexing, or binning. In general, one wishes the subset (bin) of enrollment templates returned by the pre-selection algorithm to be as small as possible. This is known as the *penetration rate*, and is defined as the average number of results returned by the pre-selection algorithm as proportion of the total database size. A *pre-selection error* occurs when a genuine template is missed by the pre-selection algorithm. The *pre-selection error rate* is the proportion of inputs with a true match enrolled in the database for which a pre-selection error occurs.

There is a trade-off between the penetration rate and the pre-selection error rate. As the penetration rate increases (i.e. fewer enrollments are matched against), the pre-selection error rate increases, as there is a greater likelihood that the true match has been omitted.

7.5 Conclusion

A wide range of performance measures for biometric systems have been presented and discussed in this chapter. A key distinction has been made between verification and identification. The identification problem is much more difficult as it is based on a series of verifications, and a failure of any may cause an incorrect identification. Due to this fundamental difference, the manner in which results are presented for verification systems differs from identification systems.

Whenever conducting an analysis, there is always a degree of uncertainty concerning the results. For example, one would be correct in having little confidence in the results from a trial that involved only 10 people. There are two issues related to this uncertainty. The first is concerned with quantifying and conveying this information. A result should never be stated as a fact, but should be qualified by a measure of variance. This is usually accomplished using confidence intervals, which specify a range where the true result is likely to lie. However, computing confidence intervals is a thorny issue, and remains an important area of research. Boxplots are another method for presenting the variability of a result. The second issue related to uncertainty is estimating the number of people necessary for a trial to have statistically significant results. There are several heuristics that give a rough indication of numbers, such as the Rule of 3 and the Rule of 30. However, this is another area which will undoubtedly benefit from future research.

Verification and identification errors are not the only potential sources of errors for a biometric system. Enrollment and acquisition errors are also possible, and have important implications for system performance. If a person cannot enroll, they will not be able to ever use the system. Therefore, investigating enrollment and acquisition errors is a necessary component of all evaluations. Similarly, template pre-selection may be used to boost the speed of an identification query. However, if the pre-selection algorithm makes an error, this will also impact system performance.

Now that a thorough examination into system-level analysis has been completed, in the next chapter the importance of user-level analysis will be demonstrated. An individual's appearance and behavior can impact their success within a biometric system, and a single person can have a significant impact on system-wide error rates.

References

[1] Agresti, A., Coull, B.: Approximate is better than 'exact' for interval estimation of binomial proportions. The American Statistician **52**, 119–126 (1998)

[2] Bolle, R.M., Connell, J.H., Pankanti, S., Ratha, N.K., Senior, A.W.: The relation between the ROC curve and the CMC. In: AUTOID '05: Proceedings of the Fourth IEEE Workshop on Automatic Identification Advanced Technologies, pp. 15–20 (2005)

[3] Bolle, R.M., Ratha, N.K., Pankanti, S.: Error analysis of pattern recognition systems: the subsets bootstrap. Comput. Vis. Image Underst. **93**(1), 1–33 (2004)

[4] Brown, L., Cai, T., DasGupta, A.: Interval estimation for a binomial proportion. Statistical Science **16**(2), 101–133 (2001)

[5] Cappelli, R., Maio, D., Maltoni, D.: Indexing fingerprint databases for efficient 1:N matching. In: Proceedings of ICARCV2000 (2000)

[6] Clopper, C., Pearson, S.: The use of confidence or fiducial limits illustrated in the case of the binomial. Biometrika **26**, 404–413 (1934)

[7] Dass, S.C., Zhu, Y., Jain, A.: Validating a biometric authentication system: Sample size requirements. IEEE Trans. Pattern Anal. Mach. Intell. **28**(12), 1902–1319 (2006)

[8] Fawcett, T.: ROC graphs: Notes and practical considerations for researchers (2004)

[9] Grother, P.J., Phillips, P.J.: Models of large population recognition performance. In: CVPR04, pp. II: 68–75 (2004)

[10] Hube, J.: Using biometric verification to estimate identification performance. In: Proceedings of 2006 Biometrics Symposium (2006)

[11] ISO: Information technology – biometric performance testing and reporting – part 1: Principles and framework (ISO/IEC 19795-1:2006) (2006)

[12] Johnson, A., Sun, J., Bobick, A.: Predicting large population data cumulative match characteristic performance from small population data. In: Proceedings of AVBPA 2003 (2003)

[13] Jovanovic, B., Levy, P.: A look at the rule of three. The American Statistician **51**(2), 137–139 (1997)

[14] Maio, D., Maltoni, D., Cappelli, R., Wayman, J.L., Jain, A.K.: FVC2000: Fingerprint verification competition. IEEE Trans. Pattern Anal. Mach. Intell. **24**(3), 402–412 (2002)

[15] Mansfield, A.J., Wayman, J.L.: Best practices in testing and reporting performance of biometric devices, v2.01. Tech. Rep. NPL Report CMSC 14/02, National Physical Laboratory (2002)

[16] Martin, A., Doddington, G., Kamm, T., Ordowski, M., Przybocki, M.: The DET curve in assessment of detection task performance. In: Proc. Eurospeech '97, pp. 1895–1898. Rhodes, Greece (1997)

[17] Payton, M.E., Greenstone, M.H., Schenker, N.: Overlapping confidence intervals or standard error intervals: What do they mean in terms of statistical significance? Journal of Insect Science **3**(34) (2003)

[18] Phillips, P.J., Grother, P., Micheals, R., Blackburn, D., Tabassi, E., Bone, J.: FRVT 2002 evaluation report. Tech. Rep. NISTIR 6965 (2003)

[19] Phillips, P.J., Scruggs, W.T., O'Toole, A.J., Flynn, P.J., Bowyer, K.W., Schott, C.L., Sharpe, M.: FRVT 2006 and ice 2006 large-scale results. Tech. Rep. NISTIR 7408, National Institute of Standards and Technology (????)

[20] Reynolds, D.A., Doddington, G.R., Przybocki, M.A., Martin, A.F.: The NIST speaker recognition evaluation - overview methodology, systems, results, perspective. Speech Commun. **31**(2-3), 225–254 (2000)

[21] Schuckers, M.: Estimation and sample size calculations for correlated binary error rates of biometric identification rates. Proceedings of the American Statistical Association: Biometrics Section (2003)

[22] Schuckers, M.: Using the beta-binomial distribution to assess performance of a biometric identification device. International Journal of Image and Graphics **3**(3) (2003)

Chapter 8
Individual Evaluation: The Biometric Menagerie

When one considers how widely the appearance of a biometric can vary, along with all the idiosyncratic behaviors that govern a user's interaction with a system, the hypothesis that all members of a population perform equally seems optimistic. However, this assumption is implicit in much of the analysis of the previous chapter, which focused on a high-level, aggregated statistics. There is no doubt that these performance measures are necessary, but it is also important to recognize that they do not tell the whole story. A system is comprised of data from a variety of sources, so collective statistics like ROC and CMC graphs necessarily preclude some information. Within any given biometric system, it is almost always the case that some users perform better than others, and it is these individual differences that are the subject of this chapter.

Section 7.1.2 introduced the concept of match score distributions for genuine and impostor matches. These distribution functions give the probabilities of genuine and impostor matches over the range of possible scores. It is a central theme of this book that these distributions are the basis for all performance analysis, and many, if not all, aspects of biometrics system have a distribution-based interpretation. For example, in Sect. 7.1.2.2 it was demonstrated that false match and false non-match rates are related to the degree of overlap between the genuine and impostor distributions. Similarly, in Sect. 7.2.1.1 the impact of database size on identification performance was illustrated through shifting rank 1 impostor score distributions. Once again, the key concept of this chapter can be posed of in terms of match score distributions.

Figure 8.1 shows how the match score distribution for the system as a whole is actually a combination the individual users distributions. Note that the individual distributions differ from person to person. For example, the distributions for User B have more overlap than for User A. This implies that User B is more likely to be involved in false matches or false non-matches. If User B is a frequent user of the system, this may have serious consequence for overall performance. This has an important implication for biometric evaluations, illustrative of another central theme of this book: there is no such thing as the "true" performance of a biometric matching algorithm. *The performance of a biometric system depends on who is using it.*

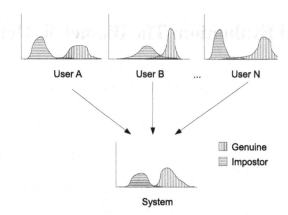

Fig. 8.1 The genuine and impostor match score distributions for a system as a whole are comprised of a combination of the distributions for individual users.

There are various ways in which a person can have poor performance. A person can be involved in a disproportionate number of enrollment failures, false matches, false non-matches, or a combination thereof. The relevance to system performance is that a small number of users may be responsible for a large number of errors. By addressing the problems experienced by a small minority, one may be able to reduce system-wide error rates significantly.

The goals of this chapter are as follows:

- Discuss individual variation and investigate the potential causes (Sect. 8.1).
- Classify typical "problem users" using the *biometric menagerie*. The biometric menagerie is a collection of animal metaphors that characterize the ways in which users can have trouble within a biometric system (Sect. 8.2).
- Introduce the *zoo plot* as a standard graph for visualizing the relative performance of individual users (Sect. 8.2.3).
- Outline a framework for the evaluation of biometric systems with a focus on user performance (Sect. 8.3).
- Present case studies for system evaluation based on the proposed user-centric approach (Sect. 8.4).

8.1 Individual Variation

Section 3.3 contains examples of poor quality data samples for a variety of specific biometrics. In this section, the discussion is more general, and is applicable to any biometric modality. The reasons why individual performance varies from person to person is discussed in Sect. 8.1.1, the impact of individual variation on system performance is presented in Sect. 8.1.2, and its relation to data quality metrics is explained in 8.1.3. Section 8.1.4 contains information about personalized match score thresholds, and in particular the limitations in their use.

8.1.1 Causes for Individual Variation

Random variation and sampling error will cause differences in the observed performance of individual users. However, differences of this kind are inevitable, and therefore of little interest. The focus of this chapter is on statistically significant differences among the user population.[1] When a real difference exists, it is assumed that there is an underlying cause. In practice, determining this cause may be complex, or even impossible. However, the general idea is that *something* is impacting a problem user's performance. If all other factors are held constant (such as equipment and environmental conditions), there are three broad categories for factors related to individual performance: the way a person is (physiology), the way they do things (behavior), and their interaction with the biometric sensors (data capture). The three categories should not be viewed as mutually exclusive, as they dynamically interact with each other.

8.1.1.1 Physiology

Within a biometric context, physiology refers to a person's physical features. The relevant part of the body depends entirely on the biometric being used. For example, in the case of a system based on retina scans it is physical properties of the eye that are of interest.

In extreme cases, an individual's biometric may be missing entirely. For example, an amputee missing both arms would be unable to use a fingerprint identification system. Therefore, they would have a 100% failure to enroll rate (albeit, in a trivial sense). Even when the biometric exists, disease or injury can damage it considerably. For example, a throat infection can severely impact a person's ability to authenticate using a speech verification system. Difficulties due to missing or damaged biometrics are unavoidable, and often insurmountable at an algorithmic level. Issues along these lines are best handled by system policy, such as using alternative, or multiple, biometric modalities (see Chap. 4).

Template aging is a well-known physiologically based challenge for biometric recognition. As people age, their biometrics change. This is a challenge, with varying degrees, for all biometric modalities.

Of particular interest are cases where a biometric is not missing, damaged, aged, or poor quality, but the owner still has performance difficulties. One example is a fingerprint that has an unusually low number of minutiae points. Minutiae embody

---

[1] Section 8.2.4.3 describes statistical tests designed to differentiate between real performance differences from those due to chance.

much of a fingerprint's distinctive information, so in theory, a population with a low-minutiae count may be fundamentally more difficult to distinguish.

In some cases, the cause of the problem is a combination between the person's physiology and the algorithms used for feature extraction and matching. For example, there is anecdotal evidence that there are certain ethnic populations in Africa who have difficulties with iris recognition systems. In this case, the issue is not related to a damaged biometric or poor quality acquisition, but rather inherent physical characteristics of the individual: the distinctive features of their irises cannot be extracted and matched using techniques that have been developed and tuned primarily using other ethnicities.

Section 9.4.1 discusses methods for dealing with situations where groups with unfavorable physiology have been identified.

Inherent "Matchability"

Except when a biometric is missing or damaged, one should be very hesitant about making the claim that a person or group's physiology makes them "inherently difficult to match". The corollary of this claim is that there are people who are *less distinctive* than the general population. This may turn out to be the case in certain circumstances (such as the African iris example given above), however more often that not physiologically-based problems can be addressed with improved feature extraction and matching algorithms. Evidence suggests that claims about people who are inherently unsuitable for biometric identification have been exaggerated in the past (see Sect. 8.5).

8.1.1.2 Behavior

For most biometric systems, an individual's behavior tends to have a greater impact on performance than their physiology. Behavior is relevant for all biometric systems because there is always some interaction between the capture device and the subject, and behavior underpins this interaction. The impact of this observation varies widely. In some circumstances, such as capturing International Civil Aviation Organization (ICAO) compliant photos for face recognition, there are strict controls in place to minimize the effects of behavior. At the other end of the spectrum, there are biometrics that are entirely behavioral, such as handwritten signatures.

The effects of behavior can manifest themselves in different ways, but the result is usually a less than optimal sample of the biometric. Some behaviors can alter the physical appearance of a biometric, while others can make it more difficult to obtain a clear capture. For example, consider a face recognition system. Smiling can change the morphology of the face, while wearing sunglasses can obstruct key distinguishing features.

Covert identification systems are an interesting case, as the subjects are unaware that they are being authenticated. In this case, any attempts to alter behavior to make it more favorable for recognition must be subtle. A discussion on dealing with behavioral-based problems can be found in Sect. 9.4.2.

8.1.1.3 Data Capture

This category is essentially a mixture between the hardware components of a biometric system and the two categories above: physiology and behavior. *The performance of a biometric system is intimately related to data capture quality, which is influenced by both physiological and behavioral factors.*

There are many ways a person's physiology can impair data capture quality. For instance, physical characteristics may hinder a person's interaction with a capture device. A crude example is a person who is too short to be properly photographed by a fixed-height camera. In other situations, the problems may be due to hardware limitations. For example, consider a person who has an unusually high voice that is outside the optimal recording range for a speech verification system's microphone. Another example is the camera settings for a face recognition system. The lighting in a face capture environment may be designed to work well for the majority of the population, but the images fail to capture sufficient contrast for individuals with very dark skin.

As with physiology, a person's behavior can affect the quality of data captures. For example, consider a fingerprint scanning device. When fingers are applied to a flat surface with too much pressure, the patterns undergo distortion, the fingerprint ridges widen, and the appearance of the minutiae may be altered (e.g. a ridge ending appears as a bifurcation). This will have a negative impact on matching performance, which relies heavily on minutiae information. A person who habitually applies too much pressure when interacting with the scanner is destined to have difficulties authenticating. For practically any biometric system, it is not difficult to imagine behavior that leads to poor quality captures.

8.1.2 Impact of Individual Variation

The previous section looked at some reasons why biometric recognition performance can vary from user to user. With these factors in mind, the hypothesis of individual variation should no longer be surprising. In fact, the null hypothesis of everyone performing identically seems unrealistic. However, the impact of user performance variation upon a system as a whole is not immediately obvious.

First and foremost, there is a direct impact on system accuracy. For example, the presence of a very small number of poor quality enrollments can have a significant impact on overall error rates. This has been observed in high-accuracy iris recognition systems where one or two poor quality iris images are responsible for almost all of the system's false matches. By simply re-enrolling these individuals, false match rates can be lowered by an order of magnitude.

Another way that individual variation impacts biometric systems is the computation of confidence intervals (see Sect. 7.3.4). Confidence intervals give the precision of an estimate for a performance measurement. The general idea is that they take into account the natural variation of match scores. Since match scores are drawn from a probability distribution, one will observe a different result every time an experiment is conducted. The amount that the result is expected to vary depends on the size of the experiment, and is quantified by a confidence interval. The early methods for computing confidence intervals assumed that the score distributions for users were identically distributed. However, the fact that the performance of individuals varies underpins the realization that the potential for variation is much greater. For example, it was mentioned above that a single person (or even a single enrollment) can be responsible for a large proportion of a system's errors. The existence or absence of a single poor quality enrollment can have a significant impact on error rates, which actually increases the uncertainty of a performance estimate. This is a core issue for research into more accurate confidence intervals.

Finally, it is likely that if one user has a particular problem within a system, there are other users having the same problem. In other words, there are likely subgroups of the population who are consistently having verification or identification problems. Detecting these user groups is the subject of the next chapter.

8.1.3 Quality Scores

Many biometric systems output a *quality score* for a sample as part of the enrollment or acquisition process. This is generally a scalar value that measures how well the algorithm was able to extract the features used for matching. In general, a template with low quality will be unsuitable for matching, and will lead to unpredictable results. Therefore, it should be rejected, and a new sample should be obtained.

A quality score is based on features inherent in the sample, and is generally designed to predict matching performance. The following are the factors that can impact the quality value:

- *Inherent qualities of the biometric*: In Sect. 8.1.1.1 the role that physiology can play in biometric recognition is examined. An ideal quality measure should reflect the amount of discriminatory information inherent in a physical biometric.
- *The fidelity of the sample obtained*: There are numerous reasons why a sample may not be an accurate reproduction of the original biometric features. Firstly, all samples suffer to some degree from random noise that is part of the acquisition process. Secondly, properties of the instrument used to obtain the sample determine how much of the original information is captured, and the precision of the measurement. For example, a high-resolution camera will capture more details than one with low-resolution. Thirdly, the presentation of the biometric may impact the quality. For example, occlusion in the image can hide important

information. Finally, environment and user behavior plays a significant role in the quality of a data capture. These factors were discussed in Sects. 8.1.1.2 and 8.1.1.3, respectively.

The physiological and behavioral factors, and their impact on data quality, are specific to the biometric modality being used. For example, a person who always applies their finger to a sensor with too much pressure will tend to have poor quality fingerprint enrollments. On the other hand, a dark room will lead to low quality images for facial recognition. The quality score measures the impact that these factors have had on the biometric sample.

A quality measure does not necessarily reflect how a sample appears to a human observer. There are some types of noise and distortions that are problematic for people, but are largely aesthetic, and can be easily removed using standard signal processing techniques. However, there will often be a strong correlation between a quality score and a person's subjective impression, as at some fundamental level both humans and machines are trying to assess the same underlying feature: the information content of a biometric sample.

Biometric Information Theory

From an information theoretic point of view, biometric information can be defined as "the decrease in uncertainty about the identity of a person due to a set of biometric features measurements" [1]. Knowing the true information content of biometric samples would be very valuable as it would formalize and quantify the fuzzy concept of "uniqueness", and provide theoretical bounds on minimum template sizes, matching accuracy performance and feasible population sizes.

Quality measures tend not to be universally applicable, even within the same biometric modality. In particular, most quality metrics are tuned to work with a specific match algorithm. The reason for this is the strong bond between quality and the feature extraction and matching components of a system. For example, a feature weighted heavily by one algorithm may not be used by another, and this should be reflected in the quality score. However, as the performance of matching algorithms approach their theoretical limits, the differences between algorithms will diminish, and quality measures will be more widely applicable as they will be a true reflection of information content.

The best way to assess the utility of a quality measure is to empirically establish its ability as a performance predictor for a matching algorithm. As a quality measure decreases, this should indicate that distinctive information is being lost. The impact of this is that the underlying genuine distributions will widen, and the mean value will decrease. Therefore, there should be a positive correlation between a sample's quality score and its mean genuine score when matched against other enrollments.

Similarly, as quality decreases, the impostor distribution will also widen. Eventually, when there is no information content in the samples remaining, the genuine and impostor distributions will be equivalent, and a system will have no ability to distinguish between genuine and impostor matches. This illustrates the importance of failure to enroll events and failure to acquire events. In order to avoid the errors inevitably associated with poor quality data, the feature extraction process should reject excessively poor quality input.

8.1.4 Individual Thresholds

A common suggestion for dealing with the differences between individual performance rates is to provide threshold levels that are specifically tailored to each person's unique genuine and impostor score distributions. Superficially, this idea seems like an attractive solution since it provides a mechanism to deal with problem users directly, and it has been demonstrated to improve system performance in laboratory settings. However, there are a number of factors that limit its use in practice:

- **Security:** The inclusion of a thresholds for each person makes the security environment difficult to monitor, as specific users could be more vulnerable to attack. For example, consider a goat who consistently receives low genuine match scores. In this case, their personalized match threshold would be set very low to reduce the probability of a false reject. However, this would make them a likely target for attack, as an intruder would only need to achieve a relatively low match score to be accepted by the system.
- **Data:** In order to make statistically valid decisions about setting a user's personalized threshold, a sufficient number of genuine match samples is necessary.[2] However, gathering this data for each potential user of a system would be time consuming and costly.
- **System complexity:** The inclusion of user specific thresholds increases the complexity of a system, thereby making it more difficult, and costly, to implement and maintain.
- **Timing:** As we age our biometrics change, so for some systems individual thresholds may need to be set up in such a way that they are continually adjusted over time.

Individual thresholds are unlikely to be useful for most large scale systems for the reasons outlined above. In general, it is better to identify the true, underlying causes for performance problems (e.g. poor enrollment quality), and address these directly. Thresholds for groups of individuals (e.g. male and female) may be of some assistance since the vulnerability is more easily managed, and the determination of a statistically significant genuine distribution is possible.

---

[2] In general, obtaining a sufficient number of impostor matches is not an issue, as the user can be matched against all other users.

8.2 The Biometric Menagerie[3]

Section 8.1 introduced the concept that individual users perform differently within a system, and discussed some of the reasons why this may be the case. For any biometric system, the distribution of match scores will naturally vary across a range of results, and it is expected that there will be some genuine matches with low scores and some impostor matches with high scores. However, a few isolated incidences of failed verifications do not warrant labeling a user a "problem user". Of interest to biometric system designers are users who *consistently* receive low genuine scores or high impostor scores, outside of what would be expected from random variation. In other words, the score distributions for these problem user groups is fundamentally different from the distribution of the general population.

There are several distinct ways in which a person may have trouble within a system. For example, a person may have trouble verifying against their own enrollment or they may repeatedly match too highly against other users. These people are familiar to biometric researchers, and have been given animals names that analogously reflect their behavior. The concept of the *biometric menagerie* was first formalized by Doddington et al. [4]. The original study was conducted on speaker verification data. However, the same concepts are applicable to all areas of biometric identification, and other studies have discovered the existence of the animals in other modalities [7, 8]. The original members of the biometric menagerie are as follows:

Sheep   Sheep make up the majority of the population of a biometric system. On average, they tend to match well against themselves and well against others. Their genuine and impostor match score distributions are consistent with the majority of other users.

Goats   Goats are subjects who are difficult to verify. They are characterized by a low genuine match score distribution, and may be involved in false rejects.

Lambs   Lambs are vulnerable to impersonation. When matched against by others, the result is relatively high match scores (i.e. high impostor distribution), leading to potential false accepts.

Wolves   Wolves are successful at impersonation, and prey upon lambs. When matched against enrolled users, they receive high match scores. In some systems wolves may cause a disproportionate number of the system's false accepts.

Each of the animals can be defined in terms of their underlying match score distributions, as illustrated in Fig. 8.2. Goats have a genuine match score distribution that is lower than most people (sheep), and lambs and wolves have particularly high impostor score distributions.

A natural question regards the relationship between these user groups. For example, if a user is known to be a lamb, does that make him or her more likely to be a goat? The traditional biometric animals are based on genuine match scores (low for

---

[3] This section is based in part on prior work [10, 11].

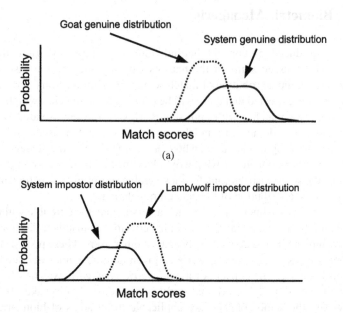

(a)

Fig. 8.2 Match score distributions for goats, lambs and wolves. (a) Genuine match score distributions for a hypothetical goat, as compared to the system (i.e. sheep) distribution. The low scores indicate a potential for false rejects. (b) Impostor match score distributions for a hypothetical lamb (or wolf), as compared to the system (sheep) distribution. The high scores indicate a potential for false accepts. In both cases, it is the relative position of the distributions that defines the animals: low genuine scores for goats, and high impostor scores for lambs and wolves.

goats) or impostor match scores (high for lambs and wolves). A new class of animals can be defined in terms of a *relationship* between genuine and impostor scores. The new additions to the biometric menagerie have combinations of low/high and genuine/impostor match scores:

Chameleons   Chameleons always appear like others, receiving high scores for matches against themselves and others.

Phantoms   Phantoms never appear similar to anyone, including themselves. Phantoms have low genuine and impostor score distributions.

Doves   Doves are the best possible users of a biometric system. The match exceedingly well against themselves, and are rarely mistaken for others.

Worms   Worms are the worst conceivable users of a biometric system, and are characterized by low genuine scores and high impostor scores. If present, worms are responsible for a disproportionate number of system errors.

Once again, the new members of the biometric menagerie can be described in terms of genuine and impostor match score distributions. However, in this case, it is the relationship between the distributions that is of interest. Figure 8.3 contains examples for the new members. For worms, the genuine and impostor distributions are close

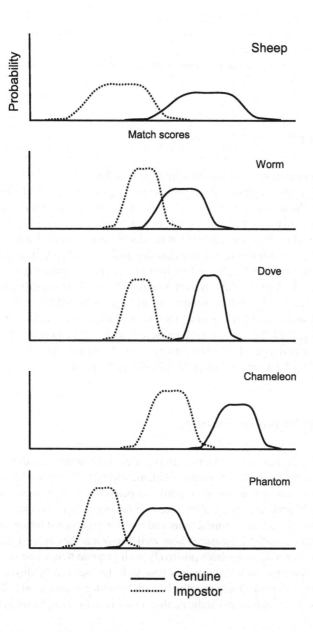

Fig. 8.3 Sample genuine and impostor distributions for a sheep, worm, dove, chameleon and phantom. The animals are defined in terms of a relationship between the genuine and impostor distributions.

together, while the dove distributions are well separated. The chameleon genuine and impostor distributions are both be relatively high, and the phantom distributions are relatively low distributions.

The following sections contain a more formal presentation of the members of the biometric menagerie.

8.2.1 Notation *

Assume a user population $\mathscr{P}$, a set of enrollment templates $\mathscr{T}$, a set of verification samples $\mathscr{S}$, and a set of matches $\mathscr{M}$. A match $m(s,t) \in \mathscr{M}$ consists of a sample $s \in \mathscr{S}$, belonging to the user $\text{person}(s) \in \mathscr{P}$, matched against an enrollment template $t \in \mathscr{T}$, belonging to a user $\text{person}(t) \in \mathscr{P}$. For each pair of users $j, k \in \mathscr{P}$, there is a set $S(j,k) \subseteq \mathscr{S}$ containing the verification results obtained by matching one of j's samples against an enrollment template belonging to k. User k's genuine scores are represented by the set $G_k = S(k,k)$, and k's impostor scores are the set $I_k = S(j,k) \cup S(k,j)$ for all $j \neq k$. A probability density function $f_S(\bullet | j, k)$ is the distribution of match scores obtained by matching samples from j against templates from k.

Each individual has two performance measures defined: one that indicates how well they match against their own enrollments, and another that indicates their ability to be distinguished from others. For a user k, let g_k be a measure of their genuine performance, and i_k be a measure of their impostor performance.

8.2.2 User Performance Statistics

The discussion so far has talked extensively of "user performance", and how it can vary from person to person. A performance statistic should reflect how well a person is performing within a system. In other words, as performance increases, the probability of a match error decreases. There are two necessary performance statistics for each user: one for genuine transactions and one for impostor transactions. The genuine performance statistic measures how well a user matches against their own enrollments, and generally correlates positively with genuine match scores. On the other hand, the impostor statistic measures how well the user can be distinguished from other people, and correlates negatively with impostor match scores. In other words, low impostor match scores indicate that a user is unlikely to be mistaken for others.

In general, any well defined measure of performance may be used, and the choice will depend on the type of biometric system under evaluation. A few possibilities are as follows:

- **Error counts:** Users can be assigned a statistic based on their actual number of verification errors within a system. Their genuine performance is based on a tabulation of their false rejects, and their impostor score is based on their number of false accepts. There are two disadvantages of this approach. Firstly, the result is dependent on a specific match score threshold. Secondly, for systems with very high accuracy rates it is likely that a majority of users are not involved in any match errors. When no errors occur, this approach has no ability to distinguish between the performance of individual users.

- **Maximum and minimum scores:** Wittman et al. [8] base a user statistic on maximum and minimum match scores. For example, the wolf score is obtained by selecting the top impostor match score for each of a user's probes. Their wolf statistic is based on the average of these top probe impostor scores.[4] The disadvantages of this approach are that it is strongly influenced by outlier scores, and is sensitive to a particular gallery. For example, the gallery may contain a person who is intrinsically similar to the probe, such as a close relation, leading to a single high impostor score. In this case, the statistic will not be representative of the user's typical performance.

- **Ranks:** For identification systems, the most relevant performance measure is how often a user is returned among the top N ranked results. In some cases, biometric systems only log the top results for an identification query. In this case, a rank-based statistic is necessary as the full range of match scores is not available.

- **Mean scores:** If a user's genuine and impostor scores are considered to be real-valued random variables, the distributions can be characterized by the expected value, which is the central moment or mean of the distributions: $g_k = \bar{G}_k$ and $i_k = -\bar{I}_k$. If a user is receiving a lot of low genuine match scores, $\bar{G}_k$ will be relatively low, indicating that the user is more likely to have trouble with the biometric system. Similarly, a lot of high impostor scores will lead to a low $-\bar{I}_k$. The advantages for this measure are that it is intuitive, robust to outliers, and places minimal restrictions on the system error rates and structure of the match score data. The primary disadvantage of this approach is that there is not necessarily a direct relationship between a user's mean score values and their participation in system errors. However, this relationship is often observed in practice, and has been empirically verified in some systems [10].

The actual statistic used to represent user performance depends on constraints of the data available, and the goal of evaluation. For the examples in this chapter, the mean score approach (i.e. $g_k = \bar{G}_k$ and $i_k = -\bar{I}_k$) is followed.

---

[4] Note that the definition of the wolf statistic is inversely related to the definition of "impostor performance" defined in this section. In other words, someone with a high wolf statistic has poor impostor performance.

8.2.3 The Zoo Plot

The *zoo plot* is visual representation of individual user performance. Each user is represented by a single point, with the x-axis plotting their genuine performance and the y-axis plotting their impostor performance.[5]

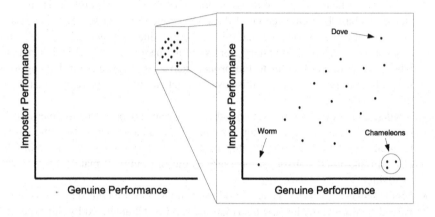

Fig. 8.4 The zoo plot. The x-axis is a measure of genuine performance (ability to authenticate), and the y-axis is a measure of impostor performance (ability to be distinguished from others). Moving in the direction towards the top-right corner of the graph indicates stronger overall performance.

The zoo plot embodies a considerable amount of information about a biometric system at a number of different levels, corresponding to the hierarchical analysis approach taken in this book:

System (Chapter 7)   In some cases, the genuine and impostor performance measures are based directly on match scores (e.g. $g_k = \bar{G}_k$ and $i_k = -\bar{I}_k$). When this is the case, the axes can plot the full range of possible match scores (see left-hand side of Fig. 8.4). With this view, the user points are a "cloud" in the upper-right region of the graph. This region represents an area where both genuine and impostor performance is high. This is a novel view of system accuracy, with the interpretation that the better the system, the closer the user "cloud" is to the top-right of the graph.

Individuals (current chapter)   As each point in the zoo plot represents a single person, the graph can be used to visually determine outliers, which are points separated from the main population. An outlier generally represents someone who is performing particularly well, or someone who is performing very poorly. In the

---

[5] Once again, keep in mind that impostor performance is inversely related to impostor scores. In other words, lower impostor scores lead to higher performance.

zoomed view of Fig. 8.4, two outliers are labeled with their animal names. The worm in the lower-left has poor genuine and impostor performance, while the dove in the upper right has high genuine and impostor performance. The zoo plot shows individual performance in relation to the rest of the population, allowing quick identification of problem users.

Groups (Chapter 9)   At a level between the whole system and individual users exists the most interesting information of all. This is where one observes correlations, clusters and trends among the population. This gives valuable insight into the nature of the population, data quality and match algorithm. For example, in the bottom right-hand corner of Fig. 8.4 there is a cluster of chameleons. By examining the properties of these three users, one may be able to determine a common cause of their high genuine and impostor match scores.

The Zoo Plot

The zoo plot is a valuable aid for viewing performance trends and quickly identifying problem users. As such, it should be considered an indispensable item in the biometric evaluation toolkit. It is complementary to the traditional graphs, such as the ROC and CMC, as the focus is not solely on overall accuracy, but also contains information regarding the performance of individuals and groups.

8.2.4 Goats, Lambs and Wolves

Goats, lambs and wolves are the original members of the biometric menagerie, and are defined in terms of genuine or impostor match score distributions.

8.2.4.1 Goats

Intuitively, goats are users who are difficult to match, and are characterized by having poor genuine performance. The definition of a goat is that their genuine match score distribution is significantly lower than those of the general population.[6] On a zoo plot, goats are found along the left-hand side (see Fig. 8.5).

Goats, like the other animals, do not necessarily represent a distinct, mutually exclusive subgroup of users [4, 7]. In fact, it is possible that they do not even exist

---

[6] The term 'goat' in the biometrics community is commonly applied to any user who has trouble authenticating in a system, regardless of the statistical significance of the effect. The difference is subtle, but one should keep in mind the distinction between the theoretical meaning of goats, and the way the term is often used in practice.

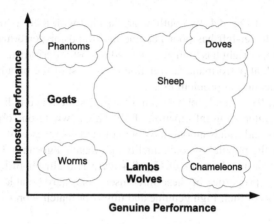

Fig. 8.5 A zoo plot labeled with regions where the animals reside.

in a particular system. The concept of goats is better thought of as a continuum, with users showing a varying degree of 'goat-like' behavior.

8.2.4.2 Lambs and Wolves

Lambs, on average, tend to produce high match scores when being matched against by other users. Similarly, wolves receive high scores when matching against others. For both of these user groups, the match score distributions are significantly higher than those of the general population. On a zoo plot, lambs and wolves are found along the bottom due to their poor impostor performance (see Fig. 8.5).

As with goats, there are not necessarily distinct lamb and wolf populations. Rather, users will display varying degrees of 'lamb-like' and 'wolfish' behavior.

Lambs vs Wolves

The definitions of lambs and wolves are symmetric. For lambs, the person of interest is being matched against (i.e. is enrolled in the system). For wolves, the person of interest is being matched against others (i.e. is being verified or identified). For many technology evaluations, there is no difference between the way verification samples and enrollments are gathered, and tests are comprised of cross-matches of all available data. In this case, there is no difference between the two user groups. However, for most real world situations there is a significant difference between verification samples and enrollment templates. For example, the enrollment procedure may use different equipment and operator intervention to ensure high quality templates. In cases such as this it is important to maintain the distinction between lambs and wolves.

8.2.4.3 Existence Test *

The definitions of goats, lambs, and wolves do not label particular users as belonging to an animal group. The definitions, in essence, simply state that match score distributions are user dependent. Once this fact is established, it follows that some users perform better than others. In this way, the presence of the animals is established without explicitly labeling users.

Hypothesis testing is used to demonstrate user dependent match score distributions. In Doddington et al. [4], the null hypothesis is formulated as follows: the density function $f_S(\bullet|k,k)$ does not depend on k. In other words, there are no significant differences between the distributions for individual users. The authors show that the null hypothesis was rejected at the 0.01 significance level using both the F-Test (analysis of variance) and the Kruskal-Wallis test for their data.

In general, the F-Test is not an appropriate method of hypothesis testing for biometric data due to its implicit assumption of normality. In reality, few distributions of biometric match scores approximate normality. Therefore, non-parametric approaches should be used where possible. One-way analysis of variance (ANOVA) is a method for testing for differences between independent distributions. Like the F-Test, ANOVA contains an implicit assumption of normality. The Kruskal-Wallis test is similar to ANOVA except that scores are replaced by ranks [3], thereby relaxing the assumption of normality. One limitation of the Kruskal-Wallis test is that it requires at least 5 random samples from each distribution.

The existence test for lambs and wolves is the same as for goats, except that impostor score distributions are used in place of genuine score distributions. The null hypothesis is that $f_S(\bullet|j,k)$ does not depend on j or k. Once again, the Kruskal-Wallis method is used to test the null hypothesis. If the null hypothesis is rejected at the 0.05 significance level, the animal groups are said to exist.

8.2.5 Worms, Chameleons, Phantoms and Doves

Goats, lambs, and wolves were defined in terms of a user's genuine or impostor match scores. The new animals differ in that they are defined in terms of a relationship between genuine and impostor performance. Unlike goats, lambs, and wolves, the specific users who belong to the new animals groups are identified. The existence test is based on whether or not there are more or less members of an animal group than expected.

Figure 8.5 contains the labeled zoo plot, which illustrates where all of the animals reside and their relation to each other. The existence of the animals is not assumed, but when they do exist there is some overlap between the original animals (goats, lambs, and wolves) and the new animals (worms, doves, chameleons, and phantoms). For example, worms would tend to be both goats and lambs/wolves.

8.2.5.1 Notation *

Let $\mathcal{G}$ be the set of genuine performance measures for all users: $\mathcal{G} = \{\cup_{k \in \mathcal{P}} g_k\}$. Rank all users $k \in \mathcal{P}$ by increasing genuine performance statistic values g_k. Let $\mathcal{G}_H \subset \mathcal{P}$ be the users whose corresponding scores are among the top 25% of $\mathcal{G}$. In other words, $\mathcal{G}_H$ is the 25% of users with the highest genuine statistics. Let $\mathcal{G}_L \subset \mathcal{P}$ be the 25% of users with the lowest genuine statistics. Similarly, let $\mathcal{I} = \{\cup_{k \in \mathcal{P}} i_k\}$, and $\mathcal{I}_H \subset \mathcal{P}$ be the 25% of users with the highest impostor statistics, and $\mathcal{I}_L \subset \mathcal{P}$ be the 25% of users with the lowest impostor statistics.

8.2.5.2 Chameleons

Intuitively, chameleons always appear similar to others, receiving high match scores for all verifications. This translates to high genuine performance and low impostor performance. Therefore, chameleons are defined by the set $\mathcal{G}_H \cap \mathcal{I}_L$.

Chameleons rarely cause false rejects, but are likely to cause false accepts. An example of a user who may be a chameleon is someone who has very generic features that are weighed heavily by the matching algorithm. In this case, he or she would receive both high genuine and impostor match scores.

In the zoo plot, chameleons exist in the lower-right corner (see Fig. 8.5).

8.2.5.3 Phantoms

Phantoms belong to the set $\mathcal{G}_L \cap \mathcal{I}_H$. Phantoms lead to low match scores regardless of who they are being matched against: themselves or others. Therefore, they have low genuine performance and high impostor performance. A possible cause of a phantom is someone who has poor quality enrollments, leading to low scores for all matches.

Phantoms exist in the top-left corner of the zoo plot (see Fig. 8.5).

8.2.5.4 Doves

Doves are the best possible users in biometric systems. They are defined by the set $\mathscr{G}_H \cap \mathscr{I}_H$. They are pure and recognizable, matching well against themselves, and receive low scores when matched against others.

An example of a dove is someone who has high quality enrollments, as well as distinctive features. Doves reside in the upper-right of a zoo plot (see Fig. 8.5).

8.2.5.5 Worms

Worms are the worst conceivable users of a biometric system, and belong to the set $\mathscr{G}_L \cap \mathscr{I}_L$. If present, worms are the cause of a disproportionate number of a system's errors. Worms crawl in the lower-left of a zoo plot (see Fig. 8.5).

8.2.5.6 Existence test *

Each user group is defined in terms of a relationship between genuine and impostor performance. For example, chameleons are users who tend to have high genuine match scores and high impostor match scores. Since the definitions are based on ranks and quartiles, the expected number of animals for each user type is $p \times |\mathscr{P}|$, where $p = (1/4)^2$. In other words, each user group should contain approximately $1/16^{\text{th}}$ of the total user population. This is under the assumption that membership in $\mathscr{G}_H$ or $\mathscr{G}_L$ and $\mathscr{I}_H$ or $\mathscr{I}_L$ are independent. However, if there is a relationship between genuine and impostor performance, this need not be the case. A dove population will be indicated by an unusually large number of members of the combined set of high genuine and high impostor performances (i.e. $|\mathscr{G}_H \cap \mathscr{I}_H| \gg 1/16 \times |\mathscr{P}|$).

The null hypothesis is that a user's genuine and impostor performance statistics are independent, and therefore there are approximately $1/16^{\text{th}}$ of the population belonging to each user type.

Assume we are interested in the set of chameleons $\mathscr{C}$ (the analysis is the same for all user types). Let c be the number of chameleons, $c = |\mathscr{C}|$. The null-hypothesis states that the probability of a particular person being a chameleon is $p = 1/16$. Since each user is independent, this is a binomial experiment with $n = |\mathscr{P}|$ trials. The hypothesis is two-sided and non-directional. Assume that the number of observed chameleons is greater than the expected number. In order to test the null hypothesis, we calculate the probability of there being at least c chameleons. This probability can be calculated using the binomial distribution:

$$f(c;n,p) = \sum_{i=c}^{n} \binom{n}{i} p^i (1-p)^{n-i}$$

For large values of n, the binomial distribution can be approximated using a normal distribution with the expected value np and variance $np(1-p)$. Assume our desired confidence level is α. The null hypothesis is rejected if $f(c;n,p) < \alpha/2$.

Since the test is two-tailed, a symmetric argument applies if the observed number of chameleons is less than the expected number. This allows for two possibilities: the null hypothesis will be rejected if there is an significantly low or significantly high number of chameleons. In other words, we can test for a significant *absence* or *presence* of the user groups. This method of hypothesis testing is non-parametric and has low computational overhead.

8.3 Analysis Using the Biometric Menagerie

Section 8.1 presented the concept of individual performance variation, and discussed reasons why this occurs. Section 8.2 examined the various ways that this performance variation can exhibit itself, and animal names were assigned to the groups of problem users. A semantic bridge is now built between the causes of user performance variation and the existence of the animals. In other words, when an animal group is observed, we would like to know what these means for the system.

8.3.1 Goats, Lambs and Wolves

Using the formal definition of Sect. 8.2.4.3, the existence of goats, lambs, and wolves means that genuine and/or impostor score distributions vary between users, and the natural consequence of this is that some users perform better than others. Previous studies have shown that goats, lambs and wolves are very common in biometric systems [10], with the practical result that user analysis should be a standard part of any biometric evaluation.

In order to identify the goats (lambs or wolves), one simply observes the users with the greatest number of false rejects (false accepts). In many cases the underlying cause of the poor performance will be apparent by viewing the biometric samples. For example, the most common cause for goat-like performance is poor quality enrollments, and this can be diagnosed through manual observation. However, if the cause of the poor performance a combination of physiology and algorithmic weaknesses (see Sect. 8.1.1.1), it may be very difficult to infer the true cause. In this case, there may be two high quality biometric samples, and the reason for the low match score (or high match score for lambs or wolves) will not be obvious.

8.3.2 Worms, Chameleons, Phantoms and Doves

The relational animals are found using the zoo plot, where they reside in the corners. When they exist in a statistically significant sense (see Sect. 8.2.5.6), it indicates a relationship between genuine and impostor match scores. In other words, a user's probability of being falsely rejected is not independent of their probability of being falsely accepted.

The existence of the animals has some relationship to system accuracy. If an animal group exists, it indicates that there is an inequality in the performance of the user population. Therefore, by definition, some users are not performing as well as others. However, there is not a direct relationship between the existence of animal groups and error rates. In other words, it is possible to have a high-accuracy system with a chameleon population, and a low-accuracy system consisting of only sheep. Therefore, strict rules concerning the interpretation of the animals are not possible, and each system must be investigated individually.

In general, the cause for the existence of the new members of the biometric menagerie is an *interaction* among the following three factors: inherent properties of the users, data quality and the matching algorithms. A change in any one of these may impact the animal population. For example, applying the same matching algorithms to two data sets, one high quality and one low quality, of the same user population can lead to very different results. Therefore, one should never make a claim that "User X is worm". Statements along these lines must be qualified, along the lines of "User X is a worm when matched using Algorithm Y, and the data quality is Z".

The following are examples within a hypothetical face recognition system of the types of situations where animal behavior may arise, for a variety of different causes. Case studies using real data can be found in the Sect. 8.4.

8.3.2.1 Chameleons

Chameleons result when a user receives high match scores for all matches. In other words, they have strong genuine performance, but poor impostor performance. A common cause of this would be a subgroup that has features in common.

Consider a face recognition system that does not explicitly model the effect of users wearing glasses during enrollment. In this case, the algorithm may mistake the glasses for a distinctive facial feature, and match all users wearing glasses with a high score. This would lead to strong genuine performance for people who wear glasses, but poor impostor performance because they falsely match other users wearing glasses. This situation is primarily a weakness of the feature extraction and matching algorithm, but also has behavioral and data quality components. In terms of behavior, wearing glasses is something people do, and in terms of data quality, glasses may make it difficult to accurately locate the eyes.

8.3.2.2 Phantoms

Phantoms receive low match scores for all transactions. This leads to strong impostor performance, but poor genuine performance. A common cause of phantom-like performance is poor quality enrollments.

An example was previously provided for physiological properties that can lead to poor quality enrollments. In some cases, very dark skin can cause images to lack sufficient contrast under normal lighting conditions. In this case, the enrollment templates contain very few distinguishing features, leading to low scores for all matches. The underlying cause for the poor performance is primarily an interaction between user physiology and the data capture system.

8.3.2.3 Doves

Doves are the best possible users in a biometric system as they have strong genuine and impostor performance. Doves generally have physiology and behavior that leads to high quality enrollments, and physical features that are easily extracted and distinguished by the matching algorithms.

For face recognition, dove-like behavior can arise from someone who has very prominent and distinctive features, such as a rare nose shape. However, having the distinctive feature alone is not sufficient, as they must also interact well with the system to create high quality enrollments. Furthermore, their distinctive feature must be extracted and weighed heavily by the matching algorithm.

8.3.2.4 Worms

Worms have poor genuine and impostor performance. In other words, they tend to receive low genuine match scores, and high impostor match scores. It is difficult to conceive of a plausible situation in which worms exist to a significant degree in a real biometric application. If there is a significant trend whereby users who score poorly against themselves score highly against others, this would likely indicate a fundamental failure of the matching algorithm.

8.4 Case Studies

The main goal of this chapter is to motivate a new approach to the analysis of biometric systems, with a focus not only on system-wide performance measures, but also with a keen eye on actual users. The primary tool for this line of analysis is the zoo plot, which shows how each individual performs in relation to the general population. The following are case studies, from a variety of biometric modalities, that illustrate how the concepts of this chapter can be used in the evaluation of real biometric data.

In all cases, average genuine and impostor match scores are used as the relevant performance measures.

8.4.1 Iris Recognition

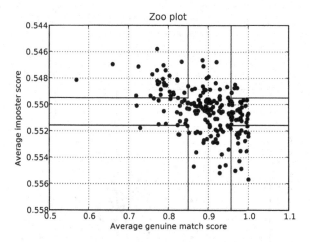

Fig. 8.6 The zoo plot for iris data. There is a distinct lack of worms (lower left) and doves (upper right), but there does exist populations of phantoms (upper left) and chameleons (lower right).

The zoo plot for an iris recognition system can be found in Fig. 8.6. In this case, there is clearly a negative correlation between genuine and impostor performance. In other words, people who have low match scores against themselves also have low match scores against others (notice that the y-axis is descending). There is a very noticeable absence of worms and doves, with only 3 or 4 of the total population (consisting of more than 200 people) falling in these regions. However, there is a significant phantom population in the upper-left corner. An analysis of these individuals shows that many of the phantoms are people who were wearing glasses when they enrolled in the system. Wearing glasses would increase the difficulty of the feature extraction task for iris recognition, leading to unreliable biometric templates that receive low match scores in many transactions.

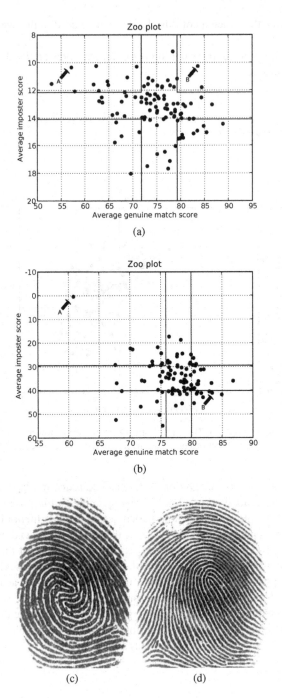

Fig. 8.7 The zoo plots for the fingerprint matching algorithms (a) *Fingerprint - Minutiae* and (b) *Fingerprint - Non-minutiae*. Both algorithms are tested on the same data set (FVC2002 DB1). The positions of two users are marked in the plots User *A* and User *B*, and sample images are presented in (c) and (d), respectively. User *A* is a phantom for both algorithm due to its unusual physiology. User *B* is a dove for the minutiae-based algorithm, but a chameleon for the non-minutiae algorithm. (c) and (d) are from the FVC 2002 DB1 [6]. ©Springer-Verlag.

8.4.2 Fingerprint Recognition

Two fingerprint matching algorithms are used for this case study: *Fingerprint - Minutiae* and *Fingerprint - Non-minutiae*. The matching algorithms differ only in the features used to calculate the match score. The fingerprint registration is performed using a two stage optimization algorithm [9]. The first algorithm uses minutiae features to calculate the similarity score, and the second uses the non-minutiae features (ridge frequency, orientation and curvature). The data set is DB1 from the FVC2002 fingerprint verification competition [6].

The zoo plots for the algorithms can be found in Fig. 8.7 (a) and 8.7 (b). Neither algorithm contained, in a statistically significant sense, any of the new members of the biometric menagerie. However, the results are still interesting.

User *A* has been indicated in both zoo plots, and has strong phantom properties for both algorithms. An example enrollment image (from FVC2002 DB1 [6]) for this person can be found in Fig. 8.7 (c). In this case, the enrollment quality of the finger is very good. There is no noticeable noise or dirt, and the ridges are clearly defined. However, compared to other fingerprints in this data set, the ridge frequency is unusually low. This low ridge frequency leads to many spurious ridges during Gabor filtering (a common pre-processing step for fingerprint images). Consequently, the minutiae data for this fingerprint is unreliable. The result of this is a low match score when matched against itself and others. This is an example of the types of situations that can lead to phantom-like behavior. In this case, the problem is due to a combination between user physiology and the feature extraction algorithm. Note that the fingerprint itself is not 'inherently hard to match'.

User *B* has a high average genuine match score for both algorithms tested. An examination of the user's images reveals that all enrollments for this finger have large capture areas, clearly defined ridge structures and consistent regions of capture. These properties make this individual easy to authenticate for both algorithms. The impostor results for this person are different between the algorithms. For the minutiae-based algorithm, the average impostor scores are low, making this user a dove. The reasons for the low impostor scores are the same as the reasons for the high genuine scores; namely, high-quality enrollments and large surface areas make it difficult for other fingers to obtain high match scores when matched against *B*. On the other hand, *B* has relatively high impostor match scores for the non-minutiae algorithm. The reason for this is that the non-minutiae algorithm is based heavily on overall ridge patterns. This fingerprint is a right-loop, which is a common ridge pattern for fingerprints, leading to high match scores when matched against many other fingers.

8.4.3 Surveillance Systems

Surveillance systems are discussed in detail in Chapter 11. The general concept is that hidden cameras are arranged to monitor movement through a designated area, called a *capture zone*. As people pass through the capture zone, their biometric is discretely recorded, and matched against a database of live targets. If a match score above a given threshold is achieved, an alarm is raised so an operator has an opportunity to investigate further.

Due to the covert nature of these systems, there is little control over the behavior of the subjects (who are unaware of their participation). Furthermore, cameras are placed in fixed positions, and cannot be adjusted to suit the physical characteristics of every person who walks through the capture zone. Therefore, the roles of physiology and behavior tend to be exaggerated in surveillance systems.

In a recent face recognition surveillance system trial there were three conspicuous animal populations:

- **Lambs:** It was observed that a very small number of individuals were responsible for a large proportion of false matches. In some cases the reason for this was not obvious. The enrollment images were of sufficient quality, and the individuals did not appear similar (to the human eye, at least) to the people they were being falsely matched against. In this case it is likely an idiosyncrasy of the feature extraction and matching algorithms.
- **Doves:** There was a distinct population of individuals who performed exceedingly well within the system. They matched against themselves consistently with high scores, and were rarely involved in false matches. In most cases, these tended to be larger individuals, who walked slowly with their head held high. This created a situation very favorable for high-quality captures. In particular, there were many opportunities for the cameras to obtain images as they lumbered through the capture zone. Furthermore, their pose increased the probability of front-on captures. This is a very clear case of how physiology and behavior can help facilitate high quality biometric samples.
- **Phantoms:** Phantoms were common in the surveillance system. In fact, there were even subgroups within the phantom population. The first group consisted of people whose clothing obscured part of their face. In particular, large sun-glasses and brimmed hats often obscured critical regions of the face. The other group of phantoms were people who walked very quickly, reducing capture opportunities, or with their head facing downwards, making it difficult to obtain a front-on capture.

8.5 Conclusion

This chapter has been concerned with biometric system accuracy from the perspective of individual users. Everyone differs in the way they look and behave. Biometric systems identify people by their physical characteristics, which are obtained through interaction with the user. Therefore, it is hardly surprising that systems will work well for some people, and poorly for others. This is an important realization as system accuracy depends not only on the matching algorithm or data quality, but on who is using the system, and how often. An implication of individual error rates is that there will be sub-populations with relatively poor performance, and this is the subject of the next chapter.

There are two important points worth mention relating to individual performance.

- Firstly, an emphasis must be placed on *relatively* poor performance. It is certainly possible, and in fact commonly observed, that even the users who have relatively poor performance still perform very well in an absolute sense. This can be stated in terms of match score distributions. There may be users whose average genuine scores are beneath the rest of the population, but as long as the genuine scores are well above the system match threshold, they are still unlikely to be falsely rejected.
- Secondly, user physiology has been shown to be a significant factor in performance. However, in most cases performance problems result from an interaction between the user's physical properties and the data capture system or matching algorithm. In other words, it is not the physical features themselves that are to blame. There is a degree of pessimism in the biometric community about people who are unsuitable for biometric identification. However, evidence is mounting that there are few people who are destined to have difficulty with biometric authentication due to inherent properties. Two recent studies support this claim.

It is often claimed that 2% of the population have fingerprints that are unsuitable for fingerprint recognition. However, an investigation by Hicklin et al. [5] of a large-scale fingerprint matching system (US-VISIT) has shown that this estimate is grossly overstated. The authors conclude that there are very few, if any, users who are intrinsically hard to match (goats and phantoms). Data quality and collection issues are the dominating factors.

The conductors of a recent iris recognition algorithm evaluation have come to a similar conclusion. IRIS06 evaluated three commercial iris recognition products in an effort to evaluate the state-of-the-art of the field [2]. The researchers found examples of wolf-like behavior, but only for specific images of a person. The authors conclude that "Doddington's Zoo phenomenon may be image-specific as opposed to individual specific". Once again, this points to enrollment quality issues, rather than inherent properties of an individual.

This is an encouraging result for the biometric research community as there does not seem to be an insurmountable obstacle of people who are inherently difficult to recognize. When errors are common for a particular individual, it can usually be addressed using improved enrollment/capture processes and robust matching algorithms.

References

[1] Adler, A., Youmaran, R., Loyka, S.: Towards a measure of biometric feature information. In Press: Pattern Analysis and Applications (2008)

[2] Authenti-Corp: IRIS06 draft final report.   http://www.authenti-corp.com/iris06/report/ (2007)

[3] Daniel, W.: Applied Nonparametric Statistics. Wadsworth Publishing Company (1989)

[4] Doddington, G., Liggett, W., Martin, A., Przybocki, M., Reynolds, D.: Sheep, goats, lambs and wolves a statistical analysis of speaker performance in the NIST 1998 speaker recognition evaluation. In: Proceedings of ICSLP-98 (1998)

[5] Hicklin, A., Watson, C., Ulery, B.: The myth of goats: How many people have fingerprints that are hard to match? Tech. Rep. NIST IR 7271, National Institute of Standards and Technology (2005)

[6] Maio, D., Maltoni, D., Cappelli, R., Wayman, J.L., Jain, A.K.: FVC2002: Second fingerprint verification competition. In: Proceedings of ICPR, vol. 3, pp. 811–814 (2002)

[7] Wayman, J.L.: Multi-finger penetration rate and ROC variability for automatic fingerprint identification systems. Tech. rep., National Biometric Test Center (1999)

[8] Wittman, M., Davis, P., Flynn, P.: Empirical studies of the existence of the biometric menagerie in the FRGC 2.0 Color Image Corpus. In: Proceedings of CVPRW (2006)

[9] Yager, N., Amin, A.: Fingerprint verification using two stage optimization. Pattern Recognition Letters **27**, 317–324 (2006)

[10] Yager, N., Dunstone, T.: The biometric menagerie. Submitted to IEEE PAMI (2007)

[11] Yager, N., Dunstone, T.: Worms, chameleons, phantoms and doves: New additions to the biometric menagerie. In: Proceedings of AutoID (2007)

Chapter 9
Group Evaluation: Data Mining for Biometrics

Chapter 7 presented methods for the holistic analysis of biometric systems, while Chap. 8 illustrated how the performance of individual users within a single system can vary. In a similar manner, certain subsets of the user population may be consistently having difficulty with the system, while others may be performing very well. For example, assume that a particular system has a significant goat population (recall that a goat is a user who has trouble matching against their own enrollments). On one hand, it is possible that each of these people has a unique reason for their poor performance. However, it is more likely that there are a few common underlying causes that affect whole groups of people. Discovering these factors, and the groups they have the greatest effect on, is an important part of the analysis of biometric systems that is often neglected.

The following are hypothetical systems that have groups of problem users:

- All fingerprints can be classified based on their overall pattern of ridges and valleys. The main classes are: left loop, right loop, whorl, arch, and tented arch. An automated fingerprint identification system (AFIS) may apply a classification algorithm to an input print, and only match it against enrolled prints belonging to the same class. This reduces the number of one-to-one comparisons that need to be conducted, reducing system load and potentially avoiding false matches. However, fingerprint classification is a difficult problem in itself, with challenges distinct from those of fingerprint recognition. Consider a system that uses a fingerprint classification algorithm for pre-selection, and further assume that the algorithm often misclassifies whorl inputs as arches. In this case, the "whorl" sub-population may consistently receive low scores, for both genuine and impostor matches, leading to a group of phantoms. In this case, it is features inherent in the physiology of the subgroup that are related to their poor system performance.
- Covert surveillance systems capture images of people without their knowledge. Therefore, unlike many biometric systems, there is very limited control over the behavior of the subjects who pass through the system. Consider a group of users who wear large sunglasses that obscure a significant portion of their face. This will hamper the ability of the face recognition algorithm to correctly identify the

individual. In this case, it is a behavioral aspect of the subgroup that leads to poor recognition performance.

- Consider a face recognition identification system that is installed at several detention centers throughout a country. At each site, detainees are enrolled in the system with a new photo, which is matched against the existing database to ensure they have not been previously enrolled under a different name. Since there are a variety of different capture locations throughout the country, the conditions at each site will vary; some variations favorable to face recognition systems, others unfavorable. For example, imagine that one site has a problem with backlighting, resulting in a disproportionate number of false accepts. In this case, the lamb population is due to environmental factors.

As these examples illustrate, there are many potential reasons why a particular group may perform poorly. Large, integrated, full-scale production systems are complex and have many sources of data. Each of these sources introduce new factors that potentially relate to system performance.

The subject of this chapter is detecting problem groups. In general, it is assumed that the biometric data is available a priori, either from an evaluation, or from a live system. Typically, the system-wide performance has already been established, and further analysis is being conducted to determine if any groups are causing a disproportionate number of system errors. Section 9.1 outlines the data relevant for this mode of analysis.

Very little research has been published about evaluating the performance of user groups for biometric systems. Traditionally, this type of analysis has been a largely manual process. The performance of common subgroups (e.g. gender and age) is established by filtering the system-level results, and computing performance measures for each group individually. This approach is explained in Sect. 9.2.

Data mining and machine learning algorithms can be used to discover patterns and trends in biometric data. This has the advantage that the process is largely automated, so in theory subtle trends may be uncovered that would otherwise be hidden among volumes of score logs and metadata. This approach is discussed in Sect. 9.3. Section 9.4 contains a discussion of approaches for dealing with problem groups once they have been identified, and Sect. 9.5 presents the limitations of group-level analysis.

After reading this chapter, you should know:

- The types of metadata information that is most relevant to group analysis (Sect. 9.1).
- How to evaluate the performance of common groups, such as age, ethnicity, and gender, by partitioning the test data (Sect. 9.2).
- How data mining techniques can be used to automatically detect groups of problem users in your data (Sect. 9.3).
- What action can be taken to deal with known problem groups (Sect 9.4).

9.1 Group Metadata

The introduction to this chapter gave several examples of user groups with poor performance for hypothetical biometric systems. The sources of the problems varied widely, from physical characteristics of the people themselves, to their behavior and to their environment. This section provides a framework for the types of factors that can impact group performance. The list is not meant to be exhaustive, as the information that is relevant depends heavily on the system under evaluation. Instead, the goal of this section is to categorize the various sources of data, and illustrate how they can impact group performance. Annex C of ISO 19795.1 contains a more comprehensive listing of performance factors relating to biometric systems [4].

It is important to note that for the purposes of this chapter, a "group" is not necessarily restricted to a group of users. A group can be defined at the template level (e.g. all enrollments from a particular location) or at the match level (e.g. all verifications that occurred in the afternoon). However, as groups of users are most commonly considered for discussion, for this chapter a "group" should be assumed to mean a group of users unless otherwise specified.

9.1.1 User Level

At the user level is information about an individual that is relatively constant over time. In other words, data that is unlikely to change between enrollment in a biometric system and subsequent verification or identification transactions. These are some examples of user level data:

- **Sex:** In some biometric systems, one gender may perform better than the other. This may be due to either physiological or behavioral reasons. For example, the vocal range of men and women is different, and may influence their performance within a speaker verification system. An example of behavioral factors are the use of fashion accessories or makeup by women that inhibit facial recognition by obscuring or altering part of the input signal.
- **Ethnicity:** Some recognition algorithms are tuned using a training set, that in essence "teaches" the algorithm to distinguish between people. However, if a particular ethnic group is over-represented in the training set, the algorithm may be biased towards them, and struggle to distinguish other groups. This is particularly relevant for face recognition systems where the effects of race are most visible, however it has been demonstrated to be a factor in other biometrics as well.
- **Occupation:** A person's occupation may, over time, alter their biometric. For example, the fingerprints of people who work extensively with their hands, such as bricklayers, are known to fade over time. This can have a negative impact on the person's performance within a system.

- **Accent:** A person's accent may be relevant to their performance within a speaker verification system.

9.1.2 Template Level

Information regarding templates and samples is the largest category because it embodies all the information relevant to the presentation of a biometric at a particular instance in time. In general, this is data that is likely to change between subsequent presentations of the same biometric characteristic. This category can be further divided into sub-categories for user, environment and system. Examples from each category are found below.

9.1.2.1 User

- **Age:** The age of the user depends on the date that the biometric was captured. In some systems, certain age ranges may pose more difficulties than others. For example, children present a unique challenge due to the fact that some biometrics (such as face and voice) can range relatively rapidly during adolescence. Other biometric systems are known to struggle with the elderly.
- **Behavior:** Certain behaviors can influence interaction with a biometric device. For example, familiarity with the system and user motivation can affect the quality of capture. However, in many circumstances behavior can be difficult to measure, quantify, or classify.
- **Mood:** Some biometrics, especially behavioral biometrics, are influenced by the mood of the subject. For example, anger can alter the way one speaks, causing a problem for speaker verification.
- **Physiology:** This category depends heavily on the biometric being used. In general, there are many physiological aspects that can change between interactions with a biometric system. For example, with face recognition an important factor is facial hair, such as beards and mustaches which can impair performance. Annex C of ISO 19795.1 contains an extensive list of physiological factors that can influence the different biometric modalities [4].
- **Clothing and accessories:** As with physiology, the relevance of clothing depends on the biometric being used. For example, large, bulky overcoats may hamper the recognition performance of gait systems, and contact lenses are known to impact the performance of iris recognition systems. On the other hand, voice and fingerprints are unlikely to be affected by clothing.

9.1.2.2 Environment

- **Time of day:** There are several reasons why the performance of a biometric system may vary throughout the day. User behavior, clothing and physiology can actually be time dependent. For example, the voice and behavior of a person who has just woken up may differ from their voice later in the day. Environmental changes, such as lighting and temperature, can also change considerably.
- **Lighting:** The lighting at the capture location can impact the quality of a biometric image. Lighting conditions are particularly relevant to face recognition systems, as they generally perform best with frontal, uniform lighting.
- **Weather:** Temperature and humidity can impact the performance of biometric systems, in particular fingerprint-based systems. Extremely hot and humid conditions can lead to sweaty finger tips, which add noise and smudging to the prints. On the other hand, cold and dry climates can lead to dry, cracked skin. Both of these situations have been known to adversely affect fingerprint-based recognition systems.

9.1.2.3 System

- **Location:** Many large-scale biometric implementations include a number of different locations where people are enrolled, verified, or identified. Each of these sites use different hardware and staff, and has unique environmental conditions. It is not uncommon to find variation between performance rates for different locations.
- **Equipment:** The equipment used to capture a biometric template or sample can influence matching performance. Some systems are designed to work with the output from a specific manufacturer. For example, an iris system may be optimized to work with images of a specific resolution, and using another camera may result in sub-optimal performance. Furthermore, faulty or dirty equipment can lead to poor quality enrollments and samples.
- **Operator:** Some systems require an operator to help users enroll, verify, or identify themselves. Variation between the operators can impact system performance. For example, a highly motivated operator may help a user achieve better results. On the other hand, an operator who neglects to clean the equipment (such as a fingerprint sensor) between presentations can be associated with poor quality enrollments.

9.1.3 Match Level

A match consists of a comparison between a sample and a template. Therefore, information at this level includes all the metadata from the user(s), template(s) and sample(s). In addition to this, it contains *relational* information. For example, consider a face recognition application in which a user enrolled while wearing glasses, but occasionally wears contacts. In this case, verification performance may depend on what they are wearing when the sample is acquired. In other words, poor performance may be caused by the *difference* between the template and sample. In theory, a relationship between any attribute pair at the template/sample level may impact performance.

The most important match level metadata concerns the length of time between the enrollment template capture and sample capture. This is known as *template aging*, and affects all types of biometric systems. Biometrics do not remain perfectly stable as one ages. Changes due to aging are particularly apparent for face recognition, but can impact any form of biometric identification. When there has been a long period of time (typically several years) between two presentations of a biometric, the recognition task is considerably more difficult. It is important to quantify this performance degradation for systems that are intended for long-term use.

9.1.4 Attribute Notation

The following notation will be used for the remained of the chapter. Assume a user population $\mathscr{P}$, a set of enrollment templates $\mathscr{T}$, a set of samples $\mathscr{S}$ and a set of matches $\mathscr{M}$. A match $m(s,t) \in \mathscr{M}$ consists of a sample $s \in \mathscr{S}$, belonging to the user $\mathrm{person}(s) \in \mathscr{P}$, matched against an enrollment template $t \in \mathscr{T}$, belonging to a user $\mathrm{person}(t) \in \mathscr{P}$. Attributes of people, templates, and matches will be represented as appropriately named mapping functions. These will not all be defined formally, but rather their meaning can be inferred from their names. Here are some examples:

- $\mathrm{sex}(p)$ gives the sex of person p
- $\mathrm{age}(\mathrm{person}(t))$ gives the age of the person contained in template t at the time of capture
- $\mathrm{score}(m)$ is the similarity score achieved by match m
- $\mathrm{quality}(s)$ is the quality score for the sample s

9.2 System Analysis Approach

The most common approach for discovering problem groups within a system is to search for them directly. This is done by segmenting the test results into subsets representing each group of interest, and analyzing each set individually. In a sense this is a *top-down* approach, as the groups are defined at a high-level (e.g. men and women), and collective system statistics are computed for each group. The analysis conducted on each subset consists of the system level evaluation techniques presented in Chap. 7.

There are several advantages to this approach. Firstly, it is intuitive and relatively straightforward to conduct. The process is clearly defined, and the results are easy to interpret. Secondly, no special data, knowledge, or software is required. All that is required is access to the original test results, and the ability to compute performance statistics.

9.2.1 Splitting the Data

The first step in the system analysis approach is to select an attribute to split on. Any of the user, template or match information from Sect. 9.1 can be used. However, the most common properties are sex, age, and ethnicity, as these represent the major subgroups for most biometric systems.

Recall from Sect. 9.1.4 that $\mathcal{M}$ designates the set of matches. The goal is to divide $\mathcal{M}$ into mutually exclusive subsets $\mathcal{M}_1, \mathcal{M}_2 .... \mathcal{M}_N$ such that $\mathcal{M}_1 \cup \mathcal{M}_2 ... \cup \mathcal{M}_N \subseteq \mathcal{M}$. Performance results are computed for each of $\mathcal{M}_1, \mathcal{M}_2 .... \mathcal{M}_N$ individually, and the results are compared to determine relative performance of the subgroups.

Assume we are interested in comparing the performance of men and women. There are three options for partitioning the test results. The samples can be filtered:

$$\mathcal{M}_{m1} = \{m(s,\cdot) \in \mathcal{M} | \text{sex}(\text{person}(s)) = \text{Male}\}$$

the templates:

$$\mathcal{M}_{m2} = \{m(\cdot,t) \in \mathcal{M} | \text{sex}(\text{person}(t)) = \text{Male}\}$$

or both the samples and templates can be filtered:

$$\mathcal{M}_{m3} = \{m(s,t) \in \mathcal{M} | \text{sex}(\text{person}(s)) = \text{Male}, \text{sex}(\text{person}(t)) = \text{Male}\}$$

$\mathcal{M}_{m1}$ includes matches for men against everyone, $\mathcal{M}_{m2}$ includes matches for everyone against men, and $\mathcal{M}_{m3}$ only contains matches of men against men. This raises an important question: when filtering match scores based on person or template attributes, should the filtering condition be applied to the samples, the templates, or both? Each approach will result in a different subset of results, and the most appropriate method depends on the goal of the test. In general, one should select the method that reflects how the system is intended to be used in a real world setting. Consider the following two scenarios:

1. A male criminal has fraudulently obtained hundreds of bank cards and their associated PINs. Assume that the ATMs for withdrawing cash are enabled with face recognition technology to verify that the person operating the machine is the rightful owner of the card. Assume the criminal tries each card with its PIN in the hope that he will generate a false accept and be permitted to conduct a transaction. Since the fraudster intends to try every card, his face will be matched against the true card owner's enrollment, regardless of their sex. In this case, $\mathcal{M}_{m1}$ is the correct test set as it contains the impostor distribution for "men against everyone", which reflects the scenario under consideration.

2. A passport issuing authority uses face recognition to ensure that an applicant does not already have a passport under a different name. The photo of the appli-

cant is only matched against other people of the same sex so that obvious false matches are not considered. In this case $\mathcal{M}_{m3}$ is the appropriate test set for determining male performance because only matches between men and men will be conducted.

In general, the question is an important one, and the answer will depend on the nature of the system being evaluated.

9.2.2 Comparing Results

After the filtering has been completed, performance statistics are computed for each set, and the results are compared. The method of comparison depends on the type of system being evaluated. For a verification system, the most common method of comparing subgroups is to plot ROC (or DET) curves for each group on the same graph. Section 7.1.3.3 contains information about the interpretation of ROC curves, which is especially useful for comparison. For closed-set identification, CMC curves can be plotted on the same graph. In this case, the superior performance is indicated by a higher identification rate at a given rank. Rank 1 graphs are another useful method for comparing subgroups. Subgroups within open-set identification systems are typically compared using alarm curves (see Sect. 7.2.2.4).

As outlined in Sect. 7.3.3, it is important to generate confidence intervals when computing results that will be compared to each other. The reason for this is to ensure that perceived performance differences actually reflect real trends, and are not due to sampling error (i.e. random chance).

9.3 Data Mining

Data mining is the process of searching through large volumes of data in an effort to discover patterns, trends, and relationships. Data mining is an umbrella term, and refers to a wide variety of processes and algorithms for knowledge discovery. The potential value of this in the context of biometrics is obvious. In theory, these techniques can automatically uncover hidden trends within a system, allowing researchers and system integrators to identify, diagnose and correct problems.

Data mining is a broad area, and there has been little work published on its use for biometric data. Two techniques for extracting knowledge will be discussed in this section. The first is a simple statistical technique that looks for relationships between attributes and performance measures (Sect. 9.3.1). The second approach is machine learning, which automatically finds patterns and relationships in the data (Sects. 9.3.2-9.3.4).

9.3.1 Correlation Analysis

Biometric systems are probabilistic by nature, due to the inherent variability of biometric samples. In other words, no two presentations of a biometric will ever be identical, so 100% certainty about a particular match is theoretically impossible. Even for powerful matching algorithms, there will be signal noise when taking digital measurements of the physical environment, leading to some uncertainty in a result. However, the key idea of this chapter is that there are some sources of variation that are intimately related to performance, and can be observed and controlled. An example of this is a *relationship* between user age and enrollment quality, where elderly people tend to have poor quality templates.

A simple approach to finding relationships between attributes and performance measures is by computing their correlation coefficient. Correlation measures the strength of the linear relationship between two variables. In other words, it measures the tendency of an attribute (often known as a predictor) to vary in the same direction as the measurement of interest. If the correlation is positive, an increase in one variable indicates a likely increase in the other variable. A negative correlation indicates the two are inversely related. For the example mentioned above, a negative correlation between age and template quality would indicate that elderly people are more likely to have poor quality enrollments than young people.

The most common method for computing the correlation of two random variables is the Pearson product-moment correlation coefficient. The input is [attribute, performance measure] pairs, and the output is the correlation coefficient, which is the strength of the linear relationship. The attribute can be any available metadata (see Sect. 9.1). In the case of categorical data (e.g. sex), each category is assigned a number (e.g. Male = 0, Female = 1). The two most common performance measures for biometrics are:

- Template quality: Many feature extraction algorithms output a quality value as a result of enrollment or acquisition. For example, an image of a smudged fingerprint would likely lead to a low quality score, while a clean image with well defined ridges would result in a high quality score. In this case, correlation analysis is used to find relationships between metadata and data capture problems.
- Match scores: A correlation with genuine (impostor) match scores may help identify groups having trouble with false rejects (accepts).

The correlation coefficient ranges from -1.0 to 1.0, with -1.0 and 1.0 indicating a perfect negative and positive linear relationship respectively (i.e. all the points lie on a straight line). A coefficient of 0.0 indicates that there is no linear relationship between the variables. Generally speaking, an absolute value below 0.3 is considered to be a small degree of correlation, and an absolute value above 0.5 is large.

Statistically Significant Correlations

When computing the correlation coefficient, one can also compute a *p-value*, which is a measure of statistical significance of the result. This is important because the correlation coefficient itself can be misleading. For example, if there are only two inputs there is guaranteed to be a perfect linear relationship between them (because there is a line that connects any two points). However, this does not prove that there is a real linear relationship. In this case, the p-value will be high, and the result is not considered statistically significant. However, a major drawback of the standard Pearson product-moment p-value is that it assumes both variables are normally distributed. This is not always the case. For example, when a categorical attribute such as sex is used, the p-value can be unfounded. Therefore, the recommended approach to verifying the significance of trends discovered by correlation analysis is by using the methods outlined in Sect. 9.2.2.

9.3.2 Machine Learning

Correlation analysis examines individual attributes, so cannot be directly used to find trends involving two or more factors. On the other hand, the top-down approach presented in Sect. 9.2 was able to test the performance of groups with multiple attributes (e.g. sex and age). A disadvantage of both approaches is that one must conduct a separate test for each potential problem group. Therefore, one must know in advance which groups are likely to be having difficulties. This is fine when the problem group is a common demographic, such as "men" or "children". However, consider a system where Asian females between the age of 25 and 45 are having trouble authenticating. In order to discover this knowledge using the system analysis approach, one would need to test many different permutations of ethnicity, sex, and age. Assume that the population is categorized into 2 genders, 5 ethnic groups, and 3 age ranges. In this case, there are $2 \times 5 \times 3 = 30$ demographics. With this number of groups, a test of each is feasible given sufficient time. However, Sect. 9.1 lists many factors that may have an influence on group performance. As more factors are considered in the analysis, there is a combinatorial explosion of the number of possible groups. For example, with 12 attributes that are divided into 3 categories, there are over $3^{12} > 500,000$ groups (although most will have few, if any, members). Obviously, if one wishes to discover trends and patterns in groups characterized by more than 2 or 3 attributes, the approach of testing each possible group directly is not practical.

This combinatorial explosion is a classic problem of artificial intelligence (AI). The *feature space* (all the possible combinations of the input attributes) of the problem is prohibitively large for an exhaustive search, so "intelligent" techniques must be used to search progressively smaller sub-spaces that are likely to contain a good, although not necessarily optimal, solution. A central focus of AI, under many different guises, is developing efficient search techniques for new problem domains. One approach is known as *machine learning*, which is concerned with the development of algorithms that allow computers to dynamically "learn" patterns from previously unseen input data.

The group analysis approach of Sect. 9.2 was referred to as *top-down*. The reasoning for this was that the groups were defined in advance at a high level (e.g. men and women), and the set of all match results was partitioned accordingly. However, the machine learning approach is significantly different, and can be viewed as *bottom-up*. Each record is associated with metadata and performance measures, and knowledge is built upon this foundation by building classifiers that model the data.

In general, there are two basic approaches to machine learning algorithms of interest to biometric applications:

- **Supervised learning**: For supervised learning each input has metadata and an associated label, and the goal is to generate a function that maps the input data to the label. For biometrics, the input would be metadata for users (e.g. sex and ethnicity), templates (e.g. capture location) or matches (e.g. time of day), and each input would be assigned a performance label. The performance label is supplied by a domain expert, and is the concept that is being modeled. For example, a person may be a "lamb", a template may be a "failure to enroll" and a verification transaction could be a "false accept". Alternatively, the label can be quantitative, such as a person's average genuine match score. An example for the output of supervised learning is a function that embodies a rule along the lines of "fingerprint verifications conducted in hot, humid conditions" $\mapsto$ "potential false accept". For this application, the goal of the process is not to develop a classification algorithm to predict the performance of unseen data, but rather to use the model that has been developed to label user groups according to performance.
- **Unsupervised learning**: Unlike supervised learning, which uses a test set of labeled samples, the input to the unsupervised learning problem is unlabeled. Therefore, the goal is not only to develop a model to distinguish the groups, but also to define the number and nature of the groups themselves. Due to its unrestricted nature unsupervised learning is more difficult than supervised learning. The most common approach is clustering algorithms, such as k-Means, which automatically discover homogeneous subgroups in the population, such as groups of people with similar properties. In the context of biometric data, the input would be all of the metadata and performance labels associated with a people, templates, or matches. The output would be groups of people defined by a set of common attributes.

Both supervised and unsupervised learning techniques can be applied to biometric data, and are treated individually in the following sections.

9.3.3 Supervised Learning

Some common approaches to supervised learning are [3]:

- **Artificial neural networks:** Neural networks are connected groups of artificial neurons that were originally motivated by the computational framework of biological brains. The weights and connections between neurons are updated dynamically to adjust the relationship between the input and output data [5]. Neural networks have been successfully applied to a wide variety of problems.
- **Decision trees:** Decision tree algorithms create rooted trees that are used for classification. Each node of the tree (including the root) contains a branching rule concerning a specific attribute, and based on the outcome of this rule, one of the sub-nodes is chosen. This process continues until a leaf is reached, which will contain a label for the instance being classified [5]. For example, the root node might contain "sex", and it would lead to separate branches for "men" and "women". A leaf contains a label such as "wolf" or "failure to enroll".
- **Naive Bayes classifier:** The Naive Bayes classifier uses probability models based on Bayes' theorem [2]. The posterior probability that an input record belongs to a given label is calculated based on the conditional probabilities that its attribute values could be obtained for that label.[1] The classification rule is defined by selecting the label with the highest posterior probabilities.
- **Support Vector Machines:** Support Vector Machines (SVMs) are based on statistical learning theory. SVMs are a binary classifier that work by finding a hyperplane in the feature space that maximizes the margin between the plane and the instances of the two classes. By mapping from the original feature space to one with high dimensionality it is able to find discriminating functions even for complex data patterns [1].

These are some of the most common algorithms used for supervised learning. Other algorithms include nearest neighbor methods, genetic algorithms, and rule induction. In theory, any of these techniques can be applied to biometric data mining. However, decision trees tend to dominate data mining applications as they have several advantages over the other techniques. However, it should be kept in mind that decision trees represent one of many learning algorithms available, and are not necessarily the optimal choice for all situations.

9.3.3.1 Decision Trees

The name "decision tree" reflects the graphical representation of the classification model, with a root, branches, and leaves. Classifications are made by following a path from the root to a leaf, making a new decision at every internal node. Figure 9.1 contains an example of a decision tree. For this example, the object of classification

---

[1] The classifier is called "naive" because it assumes that the input data attributes are conditionally independent.

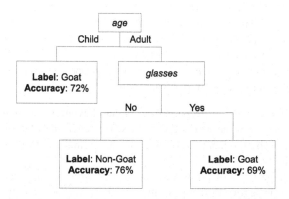

Fig. 9.1 An example of a decision tree. In this case, 72% of children and 69% of adults who wear glasses exhibit goat-like behavior (i.e. have difficulty matching against their own enrollments).

is the user's genuine performance, and they are categorized by both demographic (age) and behavioral (glasses) attributes. The resulting tree defines two goat populations: children, and adults who wear glasses. The accuracy values on the leaves are applied to all of the people who fall into that category. For example, assume there are 200 people in the example data used to build tree who are adults and do not wear glasses. 76% of these people (152) are non-goats, while the other 48 people are goats.

There are a variety of ways to build a decision tree. The most common approach is to use an information theoretic framework, which uses the concept of entropy (or information gain) in an attempt to minimize the expected number of internal nodes necessary to classify the training set [6]. The general concept is that simple trees (those with few levels) are preferred to complex trees (those with many levels) as they are more likely to reflect true patterns in the data, and less likely to be due to overfitting the data.[2] Overfitting occurs when a model has a very low error rate for training data, but is unable to generalize to unseen data.

A significant drawback of many supervised learning algorithms is their "black box" nature. In other words, there is no intuitive interpretation for the models generated for classification, which is a common requirement for data mining applications. Many methods are able to create classification rules (e.g. person X is likely to be a "worm"), however, they do not state *why* the decision was made. Decision trees are unique in that the classification decisions are easily interpreted. Other advantages of decision trees include [3]:

- Mixed data types: The input can be a mix of categorical and numerical metadata.
- Missing values: The algorithm can still generate models if some data is missing (e.g. the sex of some users is unknown).
- Robust to outliers: The presence of a few outliers will not greatly affect the tree.

---

[2] This principle is known in philosophical circles as Occam's Razor.

- Computationally efficient: Trees can be built quickly for large amounts of data.
- Irrelevant input: The trees are resistant to the presence of metadata that is not related to performance. For example, assume a fingerprint system in which ethnicity is not a relevant predictor. In this case, a tree will be built that does not include any "ethnicity" nodes. However, it should be kept in mind that the trees are resistant, not immune, to irrelevant input.

The primary disadvantage of decision trees is that they are not as powerful as some methods, such as neural networks and SVMs. However, the requirement of being able the interpret the models is important, so they remain a strong option for biometric data mining.

There is a large body of publications that deal with building decision trees. A thorough review is beyond the scope of this text, however there are two key concepts that are worth mentioning. Both concepts are related to *overfitting* the data, which occurs when trees are built that reflect idiosyncrasies of the training data, rather than general patterns within the population. These trees usually perform very well on the testing data, but do not generalize to the population as a whole. Firstly, as mentioned above, simple trees are preferable to complex trees. Large trees with many levels are usually the result of overfitting. Therefore, it is important to keep the tree relatively shallow. A technique known as *pruning* is sometimes used to "clip" branches to restrict the depth of the trees. Secondly, another technique to avoid overfitting is to randomly partition the input examples into a number of different sets, and build a tree for each set independently. These trees are then merged using an averaging technique known as *bagging* [3].

Many data mining toolkits have (such as Weka [7]) come with algorithms for pruning, bagging, and other techniques to avoid overfitting. It is strongly recommended that these techniques are used when building decision trees for biometric data mining.

9.3.3.2 Problem Formulation

There are four basic steps for the application of decision trees to biometric data. The first three of the steps are concerned solely with deciding what data will be used for the input and output of the learning task. Obviously, this is restricted by the data that is available. However, it also depends heavily on the goal of the test. The final step is running the algorithm and examining the results.

In general, it is important to have a well defined objective for a data mining task. Blindly including all information available and hoping the algorithm will sort it out is not a good strategy. When used properly learning algorithms can be very powerful, yet they are useless when used naively. One should always keep in mind the adage "garbage in, garbage out".

Step 1 - Subject: The first step is to decide the subject of classification (the first column of Table 9.1). The most common subjects will be the users of the biometric systems. In this case, the goal of data mining is to detect groups of users with

Subject of classification	Examples of input metadata	Possible labels
Templates	Glasses, contact lenses, etc.	False accept, false reject
Users	Sex, age, ethnicity	Lamb, goat, wolf
Location	Demographics, environment, etc.	Frequency of failure to enroll (FTE) and failure to acquire (FTA)

Table 9.1 Examples for the input data for the supervised learning problem. The metadata is information that may be relevant to the subject performance, and the labels are examples of categories that are applicable to the groups. For example, by using the information in the first row, one may be able to detect the properties of templates that are leading to verification errors (e.g. enrollments with glasses may lead to false accepts).

a property in common. However, learning can also be applied to other entities, depending on the system under evaluation. In general, the more complex the system, the more data available, and the greater the need to apply intelligent techniques to untangle patterns in the data. For example, individual templates can be used as the subject for classification. This would enable the discovery of trends specific to an instant in time, such as enrollment while wearing glasses. At a higher level, physical locations may be the subject of classification. A large-scale biometric system may include dozens of sites worldwide where enrollments are captured, and one may wish to know what role location plays in performance.

Step 2 - Attributes: The second step is to decide on the properties that will be used to describe the input (the second column of Table 9.1). These properties are referred to as metadata, attributes, or predictors. Obviously, the choice of attributes depends on the subject of classification chosen in the first step. The two guiding principles are to select a) the properties relevant to the subject of classification, and b) only the properties that are likely to impact performance. For example, environmental variables such as lighting may be important considerations for a face recognition location, but will have little impact on performance within a fingerprint system.

Step 3 - Labels: The third step is to define and assign the labels for the input data (the third column of Table 9.1). In essence, this step defines the overall goal of classification, and depends on the nature of the subjects. For users of biometric systems, the most common labels will be related to performance. For example, all animals from the biometric menagerie (see Chap. 8) are potential labels. These animals embody concepts like "trouble matching against their own enrollments" (goats), or "high match scores against everyone" (chameleons). The same labels can be applied to other subjects of classifications as well. For instance, animal names can be applied to locations, where a "goat" is a location where there is an unusual number of false rejections. Similarly, the animals are applicable at the template level. A "lamb" template may be an enrollment that often ranks highly against others in an identification system.

Another performance measure that may be of interest is the likelihood of enrollment and acquisition errors. Once again, these are relevant at all levels: templates, users, and locations.

The best strategy for deciding on which labels to use is to find and label two groups that have opposing meanings. For example, assume we are interested in finding people who exhibit lamb-like behavior for a surveillance system. In this case, two groups would be found: those whose average impostor rank is high (lambs) and those with low ranks (non-lambs). Samples from each group would be selected and labeled, and used as the input to the learning algorithm. The output of the algorithm would be a model that characterizes the user groups. In general, it is best to have two equal sized groups.

Step 4 - Learning: The final step is to apply the learning algorithm to generate the classification model. If a decision tree algorithm is used, each path from the root to a leaf defines a group, and associates it with a label. For each group found, one must verify that it is statistically significant (see Sect. 9.2.2).

9.3.4 Unsupervised Learning

For unsupervised learning, the input data is unlabeled, so there is no direct target for classification. Therefore, the learning process is open-ended, and there is little control over the nature of the groups found. The most common approach to unsupervised learning is based on *clustering*, which seeks to find groups that are "close" in the feature space (i.e. have a lot of properties in common). For example, applying clustering to raw biometric metadata may uncover a group of young Asian males. This indicates that they make up a distinct subgroup of the user population, but makes no implication about the performance of the group.

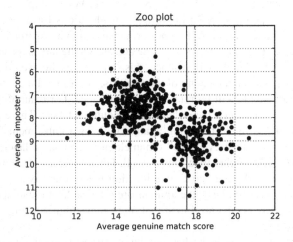

Fig. 9.2 A zoo plot with two clusters: the upper left group has low genuine and impostor scores (phantoms), and the lower right has high genuine and impostor scores (chameleons).

The problem must be formulated in a manner that emphasizes the role of performance. Instead of clustering based on metadata, clustering can be performed on match scores. For example, consider the zoo plot of Fig. 9.2. Recall that a zoo plot (see Sect. 8.2.3) plots each user based on their average genuine and impostor match score. The y-axis is reversed, so people in the upper right are performing well (high genuine and low impostor scores), and people in the bottom left are performing poorly (low genuine and high impostor scores). There appears to be two distinct clusters in Fig. 9.2, one with phantom properties (upper left) and one with chameleon properties (lower right). In theory, a clustering algorithm may be able to automatically detect these two clusters. Having found the groups, one would examine their metadata to see what properties the groups have in common. For example, it may be observed that the majority of the group in the upper left are males.

There are limitations to the clustering approach. First of all, clustering techniques work best when it is known a priori how many clusters are expected. However, in most cases this information is unknown. Secondly, groups that are nicely behaved (e.g. well separated, symmetric, and normally distributed) are easiest to detect, but real data is rarely so cooperative. Finally, in two dimensions no existing automated clustering algorithms are able to compete with the performance of the human visual system. In most cases, a quick glance at a zoo plot is more fruitful than the extensive application of unsupervised learning algorithms.

9.4 Dealing with Problem Groups

The techniques outlined in the previous sections of this chapter can be used to find problem groups within a biometric system. Typically, these will be groups of users, but can also be groups of templates (e.g. those enrolled while wearing glasses) or locations (e.g. enrollment stations X and Y). For many evaluations, the next step will be to find ways to minimize the impact of known problems.

In order to address a problem, one must first find the underlying cause. This is largely a manual process, and involves examining the data from different viewpoints. For example, assume that it has been determined that women are performing poorly in a system. In general, there are two potential underlying causes: physiological differences between men and women, and behavioral differences between men and women. In the first case, it is an inherent difference between the sexes that is causing the performance discrepancy. For example, consider the case of a speaker verification system that is better at distinguishing low pitch voices than those with a high pitch. In this case, men will tend to perform better. In the second case, a behavioral difference between men and women is impacting the ability of the system to perform a correct match. For example, consider a face recognition system that is based on skin texture analysis. In this case, makeup can cause performance problems for the recognition algorithm, impacting women more than men.

9.4.1 Physiological Problems

Problems due to between-group physiological differences are the most difficult to deal with, as they reflect weaknesses in the underlying biometric algorithm. For algorithm developers, this information is useful as it highlights areas in need of further research, and will ultimately lead to stronger algorithms. However, in many cases, the person conducting an evaluation has no access to the source code of the verification algorithms. Therefore, it may not be possible to address the root cause of the problem. The following are some possible approaches for mitigation:

- In some cases, the algorithmic adjustments will be relatively straight-forward. For example, the recognition model may be tuned using learning algorithms that can be re-trained using more appropriate test data. In some cases vendors may provide this service to their clients.
- A system policy that requires users to re-enroll periodically can be beneficial. For example, consider a system that struggles with the physiological problem of template aging. In this case, re-enrollments will reduce the problem significantly.
- Group-specific thresholds can be used to address some problems. For example, assume the children users of a system are found to be consistently receiving low genuine match scores. In this case, a lower match threshold for people younger than a certain age can be defined, reducing the risk of false rejects. However, one must be careful when implementing group-specific thresholds. In general, one must be aware of the trade-off between false rejects and false accepts, and be careful not to introduce new problems when attempting to fix old ones.
- Pre-selection algorithms can be used to avoid troublesome matches altogether. For example, if there is a high occurrence of false matches between left-loop and right-loop fingerprints, the problem can be addressed by only matching prints of the same class. Similarly, user data such as sex and age can be used to filter matches in identification systems (these are sometimes known as *soft biometrics*). While it is rarely possible to avoid the problem completely, quite often the impact can be reduced significantly.
- In some cases, a physiologically based problem may be so fundamental that new technology will be required. This may include: upgrading to the latest version of an existing engine, switching vendors, switching biometrics (e.g. from face to iris) or combining multiple biometrics. Combining multiple biometrics is known as *multimodal biometrics*, and is a powerful approach when faced with fundamental limitations of a single biometric modality (see Chap. 4).

9.4.2 Behavioral Problems

In general, problems arising from behavioral factors are easier to deal with than problems due to physiology. In most biometric systems there is some degree of control over the behavior of users, and careful procedures can be established to address the major issues found. For example, assume it has been determined that templates enrolled while the user is wearing glasses are vulnerable to false accepts by other people wearing similar glasses. A system policy that requires users to remove glasses during enrollment will eradicate the problem. In general, providing training to the users and operators of biometric systems will help lead to high quality enrollments and acquisitions, considerably reducing system errors.

One case in which there is little direct control over user behavior is covert surveillance systems (see Chap. 11). In this case, the user is not aware, and not meant to be aware, of their participation. However, even in this case there are often subtle techniques that can be used to alter behavior without the user's knowledge. For example, consider a covert face recognition system. Face recognition generally requires frontal images of subjects. However, when people are walking in everyday situations, their gaze typically wanders unpredictably around the environment. An "attractor" can be used to draw the attention of people walking through the capture zone. This is a device that has been designed to attract attention, while still looking like it naturally belongs to the environment. An example would be a brightly lit sign containing a strongly worded warning message. In this case, the surveillance cameras can be hidden behind one-way mirrors adjacent to the sign, increasing the chance of a frontal capture.

9.4.3 Environmental and Equipment Problems

There is often considerable control over the physical environment where biometrics are acquired. For example, a common problem for face recognition is lighting, and re-configuring the lighting in a room to ensure uniform, frontal lighting is a relatively straightforward task. Furthermore, the equipment used for capturing biometrics is often tightly linked to performance, and regular maintenance or replacement will lead to performance improvements.

As with behavioral problems, there are some situations in which there is limited control over environmental and equipment. For example, consider a situation where a phone-line is taped by investigators, and speaker verification is being used to identify the person talking. In this case, there is no control over the type of phone being used, and the environment in which the conversation is taking place. Similarly, consider biometric identification used at crime scenes for forensic purposes. In these situations, one must accept the limitations of the biometric data available, and rely on the development of robust techniques for pre-processing and feature extraction for accurate matching.

9.5 Limitations of Group-Level Analysis

In general, there is little research published in the scientific literature on group-level analysis for biometrics. There are a few reasons for this, some practical, and others theoretical. The researchers most likely to publish their results and findings in this area are academic. However, these groups tend to have limited access to the biometric data necessary for the task. Gathering the data from volunteer participants is time-consuming and costly. Institutions that already have access to large amounts of labeled biometric data (e.g. driver's license authorities) are unlikely to make the data available to researchers due to privacy concerns. Biometric data is inherently sensitive as it contains an indelible link to true identities.

A related factor is not just the difficulty involved in obtaining the data, but the large volume that is necessary for the analysis. The techniques of Sect. 9.2 can be used to find problems among common subgroups, such as those based on sex or ethnicity. On the other hand, the data mining techniques of Sect. 9.3 have the potential to find more subtle and complex trends among the data. However, the more subtle the trend, the more training data that is necessary to find it.

Another issue concerns the nature of the metadata attributes (see Sect. 9.1). Often the relevant attributes are difficult to measure. For example, a user's mood can be closely linked to their performance. For instance, simply being "in a hurry" can have a negative impact on performance. In other cases, such as criminal identification, users may be deliberately and actively uncooperative. However, factors such as these are difficult to quantify. They are inherently subjective, and obtaining appropriately labeled data poses significant challenges in its own right.

With all data mining techniques, there is always a risk of "over-fitting" the data. Supervised learning algorithms will almost always output *something*, even when applied to random data. One must not fall into the trap of "data fishing", where non-significant patterns are uncovered and reported. Any problem groups discovered must be verified using statistical techniques, such has hypothesis testing.

Finally, there is a fundamental limitation in that learning algorithms can only discover trends that are apparent in the metadata. However, the underlying causes for problems may not be reflected in the attributes available. In many cases, the reason why an individual performs poorly is too abstract to be predicted using the attributes such as "Caucasian" and "male". For example, consider a face recognition algorithm that performs poorly for people with a crooked nose - information about nose morphology is rarely embodied in the metadata. Ethnic labels may be useful to some degree, but certainly do not capture all inter-group variability. Therefore, underlying causes such as this are beyond the grasp of automated group level analysis, and one must rely on a visual inspections and analysis to identify sets of problem users.

9.6 Conclusion

A central theme of this book is that biometric data is complex, and performance can be evaluated at a number of different levels. This has motivated a hierarchical approach to analysis. At the highest level is system analysis, which has traditionally received the most attention, and thus reached the highest level of maturity. At the lowest level are the individual users of the system. Individual users have varying performance, and the biometric menagerie of the previous chapter is becoming recognized as an important tool for finding and characterizing these users. However, the middle layer of the pyramid, the fuzzy region between users and systems, still receives little attention. This is where groups of people, images, or locations follow patterns and trends. Unfortunately, the patterns and trends are often hidden, and require some work to identify. This chapter introduced the concept of using data mining techniques for biometric knowledge discovery. This mode of analysis is still young, and requires more research to determine the most appropriate techniques. However, due to the large volumes of data generated by biometric systems, and the importance of thorough analysis, automated and intelligent techniques will undoubtedly play an important role in the future of biometric data analysis.

References

[1] Burges, C.: A tutorial on support vector machines for pattern recognition. Data Mining and Knowledge Discovery **2**, 121–167 (1998)

[2] Duda, R., Hart, P., Stork, D.: Pattern Classification. John Wiley & Sons, Inc. (2001)

[3] Hastie, T., Tibshirani, R., Friedman, J.: The Elements of Statistical Learning. Springer (2001)

[4] ISO: Information technology – biometric performance testing and reporting – part 1: Principles and framework (ISO/IEC 19795-1:2006) (2006)

[5] Mitchell, T.: Machine Learning. McGraw-Hill Companies Inc. (1997)

[6] Quinlan, J.R.: Induction of decision trees. Machine Learning **1**(1), 81–106 (1980)

[7] WEKA: Weka data mining software, The University of Waikato. http://www.cs.waikato.ac.nz/ml/weka/ (2008)

Part III
Special Topics in Biometric Data Analysis

In Part I and Part II, the foundations have been laid for an in-depth understanding of biometric systems and biometric data analysis at a number of different levels. A few special interest topics are now examined that involve the application of these concepts to specific problem domains.

Chapters 10 and 11 look at two types of biometric systems that are currently of intense interest within the biometrics community, and public at large. Identity document systems are examined in Chap. 10, in particular how the use of biometrics can help ensure the integrity of the issuance process. This is a topical subject, as it is the first line of defense against the growing threat of identity crime. Biometric surveillance systems are designed to automatically detect persons of interest without their direct involvement or knowledge, and are the subject of Chap. 11. The interest in surveillance systems is hardly surprising, as they have the potential to play a key role in the of future law enforcement. However, there are a number of challenges for designing and evaluating these systems that are unique to the surveillance scenario.

Chapter 12 represents a significant departure from the rest of the book, in that it does not deal with the quantitative evaluation of system performance, the primary focus of the other chapters. Rather, this chapter presents a framework for evaluating a system's vulnerability to attack.

Chapter 10
Proof of Identity

Written records and verbal testimony are fundamentally fallible: documents can be forged, and people lie. Therefore, these cannot be relied upon alone to conclusively prove one's identity. This poses a difficult problem for institutions mandated with identity management. Biometrics is an appealing solution, as it provides a link between an individual and who they claim to be. This is not the panacea for all problems at hand, but will undoubtedly play an momentous role in the future of identity management and criminal investigations. This has important implications for the use of biometric data in a commercial and legal setting, which is the subject of this chapter.

This chapter is comprised of two sections:

- With the growing prevalence of identity theft, it is apparent that identity management systems, such as those pertaining to driver's license and passports, need reliable procedures for the issuance of new identity documents. The first section presents the practical ways that biometrics can be used to strengthen the issuance process, and develops novel methods for assessing the level of fraud that has infiltrated existing systems (Sect. 10.1).
- The second section considers the use of biometrics in a legal setting. This is an emerging area, and the output of biometric recognition systems is yet to reach a stage where it is fully embraced by the courts. However, there are reasons to believe that it will begin to play a greater role in future legal cases. A comprehensive treatment of the subject is not presented here, but rather a commentary on how the ideas that form the core of this book are vital to the concept of 'proving identity' (Sect. 10.2).

10.1 Identity Document Systems

Identity theft is a type of fraud that is committed by falsely assuming another person's identity, and generally involves the acquisition of ID documents under the victim's name. Once a false identity has been established, it can be used to obtain money or other benefits that the victim is entitled to. For example, if a criminal has sufficient identification under another person's name, he or she may be able to obtain credit cards or loans using the victim's line of credit. The financial and emotional damage can be severe, and it often takes a considerable amount of time and money for the victim to reestablish themselves. It is estimated that identity crime was valued at over US$55 billion in 2006, and the trend is growing worldwide. With so much money at stake, identity theft has become one of the main interests of international organized crime groups. Consequently, criminals are becoming increasingly opportunistic, devious, and sophisticated in their efforts.

The serious nature of identity crimes has exposed the key role of governments and Proof of Identity (POI) issuance agencies in strategies to prevent identity theft and fraud. The first line of defense is stricter procedures to ensure that agencies only issue identity documents to the rightful owner.

Almost all driver's license and passport documents have an associated photograph of the individual, and enormous databases of these photographs are maintained by the issuing body. A facial photograph is a biometric sample, and therefore provides a link between a photograph and the person who was photographed. Assuming that the initial acquisition of a document was legitimate, this can be used to help determine the authenticity of all subsequent applications. Therefore, there is intense interest in the use of biometric matching to contribute to a more robust POI issuance process. Furthermore, there is increasing interest in embedding additional biometrics, such as fingerprints or irises, in ID cards or passports for stronger authentication in the future.

It must be kept in mind that the aforementioned link between a photograph and the person photographed is by nature probabilistic. In other words, the matching process is not deterministic, and there is an inherent degree of uncertainty in any biometric decision (see Chap. 7). Therefore, all businesses rules and processes regarding the adoption of biometric technology must be appropriately grounded in statistical reasoning. With this in mind, Sect. 10.1.1 examines the most appropriate ways to employ the use of biometric technology. The general focus is on the use of facial biometrics by driver's license and passport authorities. However, the concepts are applicable to any organization that maintains large identity databases, regardless of the agency or biometric modality. Section 10.1.2 examines the problem of estimating levels of fraud in existing systems.

10.1.1 Modes of Operation

The primary interest in the use of biometrics in the context of identity management stems from its ability to help reduce fraud. For example, consider a typical driver's license authority. Due to its high penetration rate, driver's licenses have become a de facto "proof of identity" in many countries. Federal and state government, law enforcement, utility companies, and financial institutions rely heavily on the integrity of the driver's license for their proof of identity business processes. Therefore, improving the ability to resist and reduce fraud will have significant flow on benefits to governments and the business community at large.

There are various ways in which biometrics can be used to strengthen the POI issuance process. In general, the approaches can be divided into front of house techniques and back of house techniques. Front of house processes are conducted at the time of submission, while the applicant is on-site. A biometric decision regarding identity is made in real-time, and the result is generally presented to an operator who takes appropriate action. Back of house processes are periodically run from a central processing facility in batch mode. For example, all of the new applications for a given day may be processed overnight, with the results reviewed by dedicated investigators every morning.

10.1.1.1 Front of House

There are two basic ways that biometric matches are conducted: verification (Sect. 7.1) and identification (Sect. 7.2). Verification is a one-to-one match, and seeks to answer the question "Is this person who they say they are?". Identification involves a one-to-many match, and addresses the question "Who is this person?". In theory, either method can be used in a front of house manner when someone is applying for a new ID. However, there are limitations to the identification approach, which will be outlined below.

Verification

In the case that an ID document is being renewed, there will often be an existing photo available from the previous application. Verification can be used to help validate that the person attempting to acquire the new ID is in fact the same person who enrolled previously.

The traditional, non-biometric approach to validation is to display the previous enrollment image to an operator, who performs a visual comparison between it and the person standing in front of them. There are two problems with this approach. The first problem is known as *operator fatigue*. An operator may serve over a hundred clients in a single day, and the a priori probability of a fraudulent application is generally very small. For example, it is likely to be significantly less than 1%. Therefore, almost all of the images presented to the operator will be a genuine match. Over time, the operator will likely become tired of closely examining every image when true cases of fraud are so rare. Furthermore, they may be hesitant to inconvenience people when they are uncertain in their own mind. Eventually, they will become habituated to hit "accept" after a brief, cursory glance, adding little security to the issuance process. The second reason is that even when full attention is being paid, the verification task is very difficult. In general, and operator performance tends to be overestimated. The reality is that for most people the human visual system has a surprisingly high error rate when presented with two unfamiliar faces. Therefore, one should not rely on operators as the sole means to prevent fraudulent applications.

An obvious and natural application of biometric technology is to match a new, high-quality image of the applicant against the previous enrollment. A decision can be made based on the similarity score on whether or not to alert the operator of a potential case of fraud. In comparison with a solely operator-based approach there are two advantages. First of all, the system will operate very quickly, will be consistent, and will not be subject to fatigue. Secondly, although performance is not perfect, at least the algorithmic weaknesses can be evaluated and quantified (see Chaps. 7 - 9).

From a biometric point of view, the process is as follows. In the absence of fraud and data labeling errors, all matches will be genuine as the applicant is a correct match with himself or herself. Therefore, when a particularly low score is achieved it indicates that the applicant may not match the image on file. A match threshold is fixed, and when a verification scores below this threshold, the operator is alerted of a possible case of fraud.

In order to evaluate this scenario, the error rates of most relevance are verification performance measures of Sect. 7.1. The following are the two error rates of primary interest:

- **False non-match rate:** This is the probability that a legitimate transaction is rejected. In other words, a person is trying to renew their ID, and the system falsely suspects a case of fraud. This will occur for low scoring genuine matches. Ideally, the match threshold should be set low enough that this outcome is rare.
- **False match rate:** This is the probability that a fraudster will not be detected. In other words, a person is attempting to acquire an ID under an assumed identity, and the system does not alert the operator of a likely case of fraud. In this case, the match is an impostor match which has achieved an unusually high score. Ideally, the match threshold is set at an operating level that is high enough that impostor matches rarely exceed it.

As can be seen from these error rates, there is a trade-off between false matches and false non-matches. One of the primary goals of an evaluation is to estimate these error rates in order to help find an acceptable operating point. The ROC curve (Sect. 7.1.3.1) expresses these error rates over a range of threshold values, and is therefore an integral part of the evaluation.

Recall that error rates are based directly on genuine and impostor score distributions. The genuine distribution determines false non-match information and the impostor distribution determines false match information. The accuracy of these rates relies on the implicit assumption that the distributions were built using the same type of data that is being evaluated. Therefore, building these distributions must be done with care in order to ensure the relevance of the results.

Building a genuine distribution is relatively easy, as it only involves selecting a range of images from the same person. The participants should be selected from a random cross-section of the population. The data quality of both the enrollment image and the verification image should reflect the data quality of the real enrollment and verification images, ideally captured using the actual equipment. Another factor to take into consideration is template aging. It is important that the procedure for selecting images for the trial is not biased to select images separated by a constant length of time. Rather, a full range of time, from months to years, should separate the images.

Computing the impostor distribution generally requires more attention, and depends on the intended application. For front of house verification, the impostor distribution is used to estimate the probability that a person is not who they claim to be. However, fraudsters will not pick their victim at random. For example, a young Caucasian male would not be wise to walk into a driver's license office and try to obtain a license belonging to an elderly Asian female. This information should be made implicit in the impostor distribution by only matching people from similar demographics. In other words, building an impostor distribution by matching people at random will not accurately reflect the match scores resulting from people actually committing fraud.

Operator Reaction

The question of what action an operator should take when a likely case of fraud has been detected is important, but difficult. In general, a replacement license should not be issued on the spot. However, it should also be kept in mind that biometric matching will inevitably make mistakes, so applicants should be presumed innocent pending further investigation.

Operators and Match Scores

Another operator-related issue is the presentation of results. It is generally not advisable to present operators with raw similarity scores. This is because people without a background in biometric analysis will often invent their own interpretations of the score values. For example, people are naturally inclined to interpret any score in the range 0 to 100 as a probability estimate of a correct match. For example, a score of 99 would likely be interpreted as "there is a 99% chance that these are the same people". However, statements such as these are rarely correct. Therefore, there is a case to be made for not presenting numerical results to an operator at all. Rather, one can use a small number of well defined categories, such as "Unlikely Match", "Uncertain" and "Likely Match".

Identification

In theory, biometric matching can be used to determine if a new applicant already exists in a database under a different name. For example, a front of house system could be designed to return all likely matches for a new applicant from the database. There are two approaches for candidate selection: returning all matches above a predefined match score threshold, or always returning the top X ranked matches.

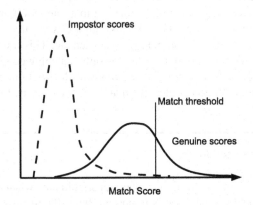

Fig. 10.1 Genuine and impostor score distributions

A threshold based approach is generally used for open-set identification, where it is unknown whether or not the query person exists in the database. This is the case for fraud detection, where actual cases of fraud are rare. Unfortunately, for large databases there will be a large number of impostor matches with high scores. This can be illustrated using the match score distributions of Fig. 10.1. First consider the impostor score distribution. A match threshold needs to be selected such that very few true impostor matches score above it. An example of such a match threshold is indicated by a vertical line. Assume that the false match rate at this point is 0.00001 (or 0.001%). For an identification query, the input is matched against every database enrollment. Consider a database with 1,000,000 images. For a given input, the expected number of false matches is $0.00001 \times 1,000,000 = 10$. Now consider the identification rate at this point, which is the area under the genuine curve, to the right of the match threshold. For the example of Fig. 10.1, the probability of a correct match looks to be around 30%. Therefore, for any given applicant, an operator would expect to see 10 false matches, and in the rare case of a fraudulent application, there would only be a 30% chance of the correct result being returned. Of course, these number depend on the specifics of the database size and quality, and the matching algorithm being used. However, these numbers are representative of what can be achieve by the best systems of today, and are unlikely to be considered adequate performance.

A rank based approach does not fare much better. Recall that a rank based approach always returns the top X ranked matches. The actual cases of fraud will be rare, so operator fatigue will be a major factor. The actual enrollments returned will be the closest matches from a very large database. In general, with a large dataset there will always be a few people who look similar to any given applicant, and this makes the recognition task for the operator very difficult. Furthermore, even if an operator is willing to look at the top 20 matches, the identification rate for large databases may still be fairly low.

In summary, with a large database there will always be people with coincidental similarities, so full-scale front of house identification is generally not feasible. This is another example of the dependence of identification performance on database size, as explored in Sect. 7.2.1.1. There are two ways to circumvent this problem. The first approach is to match applicants against a relatively small watchlist instead of the entire database. For example, the watchlist can contain several hundred persons of interest, for example people suspected to be actively involved in identity theft. In this case, a match threshold approach would work well. The second approach would be to use multiple biometrics. For example, if fingerprints were also enrolled, a front of house identification approach would be feasible, even for databases with millions of records (see Chap. 4).

10.1.1.2 Back of House

In general, when a new identity document is being requested, there is no need for it to be issued on the spot. Therefore, both of the methods presented in the previous section can also be implemented as back of house procedures, and the process is essentially the same. There are several advantages of this approach:

- The processing can be centralized. This equates to an easier and less expensive implementation, as it does not require software licenses and terminals at many locations throughout a country.
- There is a computational advantage in that the processing does not need to be conducted in real-time.
- Trained and dedicated investigators can review all cases. In this case, they have the time, expertise and resources to examine potential cases of fraud much more thoroughly than a front of house operator.

In addition to verification and identification, there are other uses for back of house biometric technology:

- Many databases have problems with data integrity, but this information is difficult to detect manually due to the sheer volume of data. Biometrics can be useful for data cleansing processes. For example, data entry errors can lead to the same person existing under multiple identities in a database. In essence, cases such as this have similar properties as cases of fraud, so similar techniques can be used to detect them. Similarly, experiments can be designed to uncover men accidentally labeled as women and vice versa, as well as other data integrity issues.
- Agencies that maintain large identity databases are often contacted by other agencies to help with their investigations. For example, consider a case where a person

has been found in a coma, and their identity is unknown. An off-line identification tool could be a service provided to other agencies to help conduct these investigations.

The primary disadvantage of back of house operations is that fraudsters are not "caught in the act", and a valuable opportunity for apprehension may be lost. However, as mentioned in the previous section, biometric decisions are rarely strong enough to justify immediate and decisive action anyway.

10.1.2 Estimating Levels of Fraud

Section 10.1.1 presented two ways in which biometrics can be used to help prevent the issuance of fraudulent identity documents. However, in most large scale identity databases, there will already be existing cases of fraud. Estimating the extent of fraud is difficult, and many organizations have no idea how common it actually is.

For our purposes, a case of fraud is defined as a single instance of a person obtaining an ID under another person's identity. Biometrics can be used to automatically detect this situation because there may be some match pairs labeled as genuine but are actually impostor (i.e. the application photo vs. the previous enrollment photo) and there may be some genuine matches labeled as impostor (i.e. in the case where the same person has committed fraud multiple times). In both cases, these match pairs will tend to exhibit themselves as outliers since they are drawn from one distribution, but labeled as the other.

There are a number of assumptions that are made with regards to the following analysis:

- People committing fraud do not actively change their appearance to a) look like their victim or b) look unlike themselves. If this is the case, it would alter the underlying match score distributions. However, the extent to which the match score distributions change would depend on the effort and skill of the fraudster, and is therefore very difficult to estimate. However, in our experience fraudsters make very little effort to alter their appearance beyond simple measures such as wearing glasses or changing their hairstyle, and a robust face matching algorithm already minimizes the impact of these factors.
- Mistakes made during the manual investigation process are not taken into account.
- The data integrity is assumed to be high. A ID labeling error in a database (i.e. accidentally assigning the wrong name to an identity document) has the same properties as a case of fraud.
- It is assumed that a fraudster has at least two enrollments in the database. One of these may be legitimate (i.e. their real driver's license or passport) or there may be a combination of legitimate and other false IDs.
- Genuine and impostor scores are independent and identically distributed, and covariance between the two is not modeled.

10.1.2.1 Running the Experiment *

The experiment is composed of two steps. In the first step, a large scale cross-match is conducted to find impostor matches with very high match scores, which are likely cases of fraud. In the second step, this list of candidates is culled by matching suspected fraudulent enrollments against other enrollments with the same name.

Step 1

The general idea of the first step is to observe random, high-scoring match pairs. Matches that score in this range are common for genuine matches, but very rare for true impostor matches. Therefore, this offers some evidence that a match is a possible case fraud.

It is generally not practical to conduct a full cross-match of a whole database if its size is in the millions, as this would lead to trillions of matches. Therefore, the database is randomly sampled to create a smaller, more manageable test set. Consider a database of enrollment images $\mathscr{D}$ with size $N = |\mathscr{D}|$. The process is as follows:

1. Randomly sample a proportion p of enrollments from $\mathscr{D}$. This will be the test set $\mathscr{T}$, and will be comprised of pN images. For example, if the database contains 1,000,000 images and a sampling rate of 10% (i.e. $p = 0.1$) is used, the test set $\mathscr{T}$ will contain 100,000 images.
2. Conduct a cross-match of all images in the test set, resulting in a set $\mathscr{M} = \mathscr{T} \times \mathscr{T}$ of matches. Exclude from $\mathscr{M}$ matches known to be between the same person. Therefore, in theory $\mathscr{M}$ only contains impostor scores. Since most matching algorithms are symmetric, it is actually only necessary to conduct half of the matches. In other words, if image A has already been matched against image B, it is not necessary to match B against A. The resulting number of match results is $|\mathscr{M}| \approx (pN)^2/2$. This is actually an upper bound, because the true number depends on how many genuine matches were excluded. However, the number of impostor matches generally far outweighs the number of genuine matches, so this is a good approximation. Using the previous example, the number of matches would be $|\mathscr{M}| \approx (0.1 \times 1,000,000)^2/2 = 5$ billion. This is a large number, but not unreasonable for today's technology.
3. Select a threshold t_1 to remove most members of $\mathscr{M}$, leaving only the highest scoring matches. In other words, the threshold is selected so that the false match rate is extremely low. For example, select t_1 such that the false match rate FMR(t_1)=1.0×10^{-6}. In other words, only 1 in 10 million true impostor scores exceed the threshold.[1] Surprisingly, the correct match rate at this threshold may be relatively high. For example, it would not be unreasonable for a matching

---

[1] In order to select t_1 at this level, it is necessary to know the impostor score distribution with a high degree of precision. This should be established experimentally using a data set with the same properties as the one under investigation.

algorithm with an EER around 3%. to have a correct match rate of 70% (i.e. FNMR(t_1)=0.3) at a false match rate of 1.0×10^{-6}.

4. The resulting set is $\mathscr{I}$, the set of highest scoring matches: $\mathscr{I} = \{m \in M \,|\, m \geq t_1\}$. Assuming no fraud, the expected size of the set is $|\mathscr{I}| = |M| \times \text{FMR}(t_1)$. In other words, this is the number of true impostor matches that would be expected to be above t_1. For the running example, the expected number of impostor matches in $\mathscr{I}$ is $5,000,000,000 \times (1.0 \times 10^{-6}) = 5000$.

Step 2

The second step is to add matches from the set $\mathscr{I}$ to a set $\mathscr{F}$ by excluding people who look similar to their previous enrollments. For each match $m(a_1, b_1) \in \mathscr{I}$, a_1 is an enrollment labeled as person a, b_1 is an enrollment labeled as person b, and $a \neq b$. We have one additional piece of information that can be used to test the likelihood that either a or b is a case of fraud. Namely, if either a_1 or b_1 is a case of fraud, they will have a low score when matched against other enrollments for a and b, respectively. We now select a second match threshold t_2 that is used to eliminate most impostors from $\mathscr{I}$.

For example, assume that $|\mathscr{I}| = 5500$, and contains 5000 actual impostor matches, and 500 actual cases of fraud. Further, assume t_2 is selected such that the false non-match rate is 10% (i.e. FNMR(t_2)=0.1). In other words, the probability that a true genuine match scores above t_2 is 90%. Consider the match $m(a_1, b_1) \in \mathscr{I}$ where both a and b have other enrollments in the database, labeled as a_0 and b_0 respectively. Let $m_a = m(a_0, a_1)$ and $m_b = m(b_0, b_1)$. If $m_a < t_2$ or $m_b < t_2$, add m to $\mathscr{F}$. In other words, if either a or b does not appear to match their other enrollment, add m to our set $\mathscr{F}$. To see the impact of this, we must look more closely at m. The match m will fall into one of two categories:

Non-fraud   Assume that m is not a case of fraud. Therefore, a and b are two different people who just happened to look like each other (according to the match engine), and a and b should produce high match scores against their previous enrollments. In other words, m_a is genuine and m_b is genuine. The probability that $m_a \geq t_2$ and $m_b \geq t_2$ is $0.9 \times 0.9 = 0.81$, and m is not added to $\mathscr{F}$. Consequently, 81% of the non-fraud cases will not be added to $\mathscr{F}$.

Fraud   Assume that m is a case of fraud. In this case, a and b are the same people, and a_1 is mislabeled, b_1 is mislabeled, or both are mislabeled. Therefore, at least one of m_a or m_b is an impostor. We are interested in the probability that $m_a \geq t_2$ and $m_b \geq t_2$, which would mean that a real case of fraud is being excluded from $\mathscr{F}$. This probability depends on the false match rate at t_2. A conservative example would be a FMR(t_2) of 0.01 at a FNMR(t_2) of 0.1. Therefore, for this example, the probability of one of m_a or m_b being an impostor and both scoring above t_2 is about 1%. If they are both impostors, the chances are even less of them both scoring above t_2.

Step 2 succeeds in excluding most non-fraud cases from $\mathscr{F}$, while adding almost all actual cases of fraud. With our running example, of the 5000 true impostors in $\mathscr{I}$, 4050 would be not be added to $\mathscr{F}$, while 495 of the 500 actual cases of fraud would be added to $\mathscr{F}$. In summary, from our original hypothetical database of 1,000,000 records, and the subsequent 5 billion matches, we are left with a set $\mathscr{F}$ of approximately 1000 match results, about half of which are likely to be cases of fraud.

10.1.2.2 Manual Investigation

The goal of manual investigation is separate the true impostors from the frauds in set $\mathscr{F}$. Some of the matches will be obvious impostors, and the reason why they achieved such a high score will be due to idiosyncrasies of the matching algorithm. These matches can be quickly discarded, as they are inconsequential. In other cases, it will be obvious that the images contain the same person, and there will be an obvious reason why. For example, it may be due to a data entry error. However, at the end of the day, the process is bound to be difficult and manually intensive. In our experience, with such a large set of data under investigation, every conceivable boundary condition is likely to occur. For example:

- In the case of identical twins, there will be two people who have an almost identical biometric sample (assuming facial photographs are being used), but it is not fraud. For men and unmarried women, this can often be resolved simply using surnames. However, in other cases, it may be more difficult.[2]
- There are many cases where people can legitimately have duplicate IDs with different names. For example, undercover operatives may have multiple identities. In cases such as this, care must be taken not to compromise the safety and security of the undercover operative. Providing investigators with a "black list" of people to exclude from fraud investigations is in itself a major risk. There is no easy way to deal with this situation. A good option is that cases such as these are not processed and stored as regular IDs, but are issued through a special procedure with its own independent, and secure, records.

A significant portion of $\mathscr{F}$ will be impostor matches containing people who simply look very similar, and cannot be distinguished conclusively by inspection alone. Before labeling a match a case of fraud, additional investigation must be conducted. For example, one may look at the supporting documentation that was provided at the time of application. This will often have strong clues (such as handwriting style, previous address information, etc.) as to the authenticity of the applications. Most

---

[2] We are aware of a case in which identical twin brothers had the same first and last name, and only differed by their middle initial. This caused so much confusion at the driver's license authority that investigators required the brothers to both show up at their offices in person at the same time in order to confirm that they were in fact two different people. A photograph of them together, as well as a signed statement, was recorded for future reference.

driver's license and passport authorities will already have a fraud department with the resources and skills to conduct this investigation.

10.1.2.3 Estimating Fraud Levels *

After the completion of the manual investigation, a set of samples is left with true cases of fraud. The final step is to use this information to infer underlying fraud rates. A high-level outline is presented below.

First the following question is addressed: given a random person who has two or more IDs with different names, what is the probability that they will be detected in the experiment? Let p be the sampling proportion, i.e. the proportion of the database $\mathscr{D}$ that was included in the test set $\mathscr{T}$. Assume that a person has two enrollments in the database under different names. The probability that both of these images will be included in the test set $\mathscr{T}$ is $p \times p$. We are now interested in the probability that this match will be included in the set $\mathscr{I}$, which is created by filtering $\mathscr{M}$ to retain only high scoring matches. The fraudulent match pair is drawn from the genuine distribution, so we are interested in the probability of a correct match, which is 1.0-FNMR(t_1). Finally, we need to know the probability that the match is added to the set $\mathscr{F}$. This is the probability that an impostor matched against the previous enrollment match scores below t_2, or 1.0-FMR(t_2). Thus, the final probability is as follows:

$$\text{probability of detection} = p \times p \times (1.0 - \text{FNMR}(t_1))(1.0 - \text{FMR}(t_2))$$

For the example of the previous sections, we have $p = 0.1$, FNMR$(t_1) = 0.3$, and FMR$(t_2) = 0.01$. Therefore, the chance of detecting a case where an individual has two images with different names in the same database is about 0.7%. Based on this value and the number of cases actually detected, one can estimate the actual level fraud. For example, if 5 true cases are detected, we can estimate that there are approximately $5/0.007 \approx 715$ cases in the full database.[3]

As it turns out, as people commit more fraud, the probability of their detection increases rapidly. This is beneficial, as many fraudsters are prolific. Once they learn the process to obtain fraudulent IDs, they are likely to do it many times. In fact, several cases in which people have obtained over 100 instances of fraudulent ID from a single issuance authority have been seen. The impact of this on detection probabilities is dramatic, because their chance of detection grows quadratically. This is illustrated in Fig. 10.2, which shows how the probability of detection increases with the level of fraud committed. For this example, people with more than 30 enrollments are almost certain to be detected, despite the fact that they are buried in a database with a million enrollments.

In summary, the proposed technique is not very useful as a method for exhaustively finding all existing cases of fraud in a database. In fact, it is unlikely to find

---

[3] The analysis is more complicated when the more general case is modeled where fraudsters have an arbitrary number of fraudulent IDs in the system.

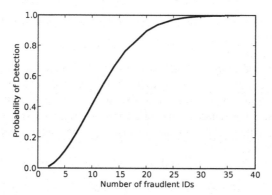

Fig. 10.2 The probability that a random fraudster is detected during an experiment, based on their number of fraudulent IDs. The graph is based on a hypothetical database and matching algorithm, using a sampling rate of 10%, an operating point t_1 with a correct match rate of 70%, and operating point t_2 with a correct non-match rate of 99%. The x-axis measures the number of fraudulent licenses held by the fraudster, and the y-axis is the probability of their detection. For example, for a person who holds 10 licenses, there is approximately a 40% chance that at least one fraudulent pair will be included in the experiment results.

any given person who has only two or three fake IDs. However, the probability of detection increases very quickly with the corresponding rate of fraud. Therefore, it can be used to find most of the habitual offenders. Furthermore, through careful reasoning, the likelihood that any given case of fraud will be detected can be determined. In turn, this can be used to estimate underlying fraud levels. However, it should be kept in mind that the procedure is based on a number of assumptions and estimations, so any reported outcomes have a high degree of statistical uncertainty.

10.2 Using Biometrics as Legal Evidence

In many criminal cases, in order to prove that a suspect is in some way associated with a crime, it is first necessary to prove that two images, video clips, or sound recordings contain samples from the same person. The proper use of biometrics in a legal settings is a broad subject, and could easily fill up a volume of its own. However, it is worth noting that many of the ideas and techniques presented in this book are fundamentally relevant.

Human fingerprint recognition for forensic purposes is well established, and has been used in legal proceedings for over a century [1]. More recently, there is some precedence for the use of forensic speaker identification [2]. Traditionally, the approach has been to rely on human *expert witnesses* who base a decision on their experience. For example, they may say something along the lines of "In all my 20 years in the field, I've never seen a match that so clearly ...". This form of argument is known as an *appeal to authority*. The defense or prosecution is basically claiming that the witness is a recognized expert in the field, and therefore should be trusted. On one hand, the witness will often be correct. However, on the other hand the intuitions of an expert are not a suitable replacement for a rigorous, scientific, and probability-based argument.

Expert witness testimony is being increasingly challenged in courts [1]. People are inherently biased, often without realizing it. For example, expert witnesses may be subconsciously motivated to support a particular outcome, depending on who is paying for their time. Furthermore, they often have an implicit assumption of guilt or innocence, which will invariably impact the manner in which they read and interpret the available evidence.

10.2.1 Advantages of Biometrics

The field of biometrics offers an attractive alternative to expert witnesses for several reasons:

- A single piece of evidence is never conclusive - its purpose is to add weight to a given hypothesis or its alternative. In theory, a decision is reached "beyond reasonable doubt" by a jury who judiciously considers all evidence available. Therefore, each piece of evidence should have an associated probability value. Expert witnesses may provide a probability estimate, but these are rarely based on empirical studies, and tend to be justified with a great deal of hand waving. A primary advantage of biometrics is that the field is fundamentally grounded in statistical reasoning (see Chap. 7). Therefore, biometric "decisions" come with a probability estimate (based on the underlying genuine and impostor match score distributions) and a measure of statistical uncertainty (based on the size of the experiment).
- Biometric matching algorithms do not hold personal prejudices. This is not to say they are unbiased or that the results cannot be misinterpreted or misused, but there is a degree of subjectivity and impartiality that is difficult to achieve with human testimony.
- Biometric authentication offers a high degree of transparency, and is open to scrutiny by both defense and prosecution. Algorithms will inevitably have inherent biases due to their design and internal model. However, where these biases exist, they can be uncovered through a process of empirical evaluation. Once again, this is generally not possible with expert witnesses.
- Biometric algorithms have no memory. Therefore, it is possible to use them to test a series of different scenarios and hypotheses, with the outcome of one not influencing another.
- Biometric matching is very fast, allowing one to conduct large scale experiments that would not be feasible for human subjects. For example, to test an algorithm's ability to distinguish images of a particular population demographic, tests can easily be run that include thousands of individuals and millions of matches.

10.2.2 Disadvantages of Biometrics

Despite the advantages, it should be emphasized that computers are not infallible. People may be inclined to put blind faith in the output of an algorithm due to a feeling that computers are deterministic, and do not make mistakes. This is entirely wrong.

In fact, the title of this chapter, "Proof of Identity", is a misnomer because no proof is actually offered, at least in a *deductive* sense. In reality, biometric verification is based on probability distributions, and therefore the "proofs" are better thought of as *inductive*. In other words, a verification decision is based on reasoning along the lines of: "this is probably a genuine match because previous genuine matches I've seen tend to score in this range". Furthermore, the matching process itself is based on an underlying model, which makes assumptions of its own, and may be biased in unpredictable ways.

One of the reasons that courts are hesitant to accept the outcome of automated decision making processes is that people are uneasy with the idea that a legal outcome (which may ultimately be a person's freedom, or even life) may be falsely determined by a programming or processing error. This is a valid point and deserves careful consideration. However, it should also be kept in mind that human error has the potential to be just as costly, and is certainly not uncommon.

10.2.3 Match Score Distributions

One of the major themes of this book is that match score distributions lay at the heart of all biometric system. Correspondingly, almost all concepts in biometrics can be interpreted in terms of genuine and impostor score distributions. Another major theme is that these distributions do not exist independent of the context in which they are being used. In other words, a verification algorithm does not have a "true" error rate that is universally applicable across the board. An algorithm that works well in some circumstances will invariably work poorly in others. The most important factors are data quality and the user population. This has profound implications for the manner in which testing should be conducted when building a proof of identity argument:

Data quality   The accuracy of a matching algorithm is heavily dependent on the quality of the biometric data. For instance, imagine a grainy CCTV image has been obtained from a crime scene, and face recognition is being used to help establish the identity of the suspect. In this case, genuine and impostor distributions (and subsequent error rates) derived from sharp, well-lit, full-frontal, high-resolution images are completely irrelevant. Therefore, every effort must be made to collect data of a similar quality as the samples under consideration.

User population   The population that should be used for the tests is another important issue. In particular, consider the following question: should a random sample from the entire population be considered, or should one be more selective? The resolution recommend is that the genuine and impostor distributions be established separately. For genuine match scores, other individuals from the same demographic should be selected. This will answer the question "how well is the algorithm able to confirm the identity for people with similar physical characteristics as the suspect?". For instance, if the suspect is a young Caucasian male, the ability of the algorithm to verify the identity of an elderly Indian woman bears no consequence. For the impostor distribution, the population should be selected from the the full range of likely suspects. For example, consider a voice recording that is being used for identification. It will normally be obvious if the speaker is male or female. Assuming it is male, the impostor scores should be sampled from a variety of other males. In essence, this answers the question "how well is the algorithm able to distinguish this suspect from other *possible* suspects?". Including samples from the entire population makes the recognition task too easy, and in this setting one should always err on the side of caution.

Sample size is another important consideration. It is vital to be able to ensure statistical significance of the results, and this is done by including a sufficient number of samples in the test set. These topics are covered in detail in Sect. 7.3.

Another important consideration is the manner in which the results are presented to the jury. First and foremost, one should not attempt to present the matching process as a mysterious black box that never makes mistakes. One of the advantages of using biometrics is that, when used properly, it opens the lid of the black box and allows everyone to peer in. Most core concepts behind biometric matching can be explained intuitively to general audience (see Part I of this book), and every effort should be made to do so.

10.3 Conclusion

The focus of this chapter has been on using biometrics as a technique to help "prove" identity, in both business and legal contexts.

As seen in Sect. 10.1, biometric matching can provide a valuable contribution to improving POI procedures, as it provides a direct, ideally non-alterable, link to a person's physical identity. Having stronger identity management systems has significant benefits not only to government, but also to businesses and the wider community. With the threat and serious implications of identity crime in mind, biometrics should be considered a mandatory component of any future identity system.

Section 10.2 discussed the application of biometrics in a legal setting. It is not expected that the adoption of biometrics will occur overnight, nor is it likely that they will ever completely replace human testimony. However, we feel strongly that they have much to offer in the way of a scientific tool that helps establish identity. DNA evidence is used extensively court, and is often presented as the ultimate and infallible identifier. However, DNA matching is fundamentally based on probabilistic matching techniques. Furthermore, the matching process is largely automated and relies heavily on the use of sophisticated technology. In this sense, it mirrors the biometric matching process very closely. Therefore, there are no insurmountable barriers to the acceptance of biometrics, and considering the potential benefits, it seems likely that their use will gain momentum in the near future.

References

[1] Cole, S.A.: History of fingerprint pattern recognition. In: Ratha, N., Bolle, R. (eds.) Automatic Fingerprint Recognition Systems, pp. 1–25. Springer (2004)
[2] Rose, P.: Forensic Speaker Identification. CRC Press (2002)

Chapter 11
Covert Surveillance Systems

Covert biometric surveillance systems are used to identify people without their knowledge. There are many possible applications of this technology, with the most prominent being related to law enforcement. Consider the following scenario.

A bomb has detonated at a busy train station during peak hour, causing scores of injuries and deaths. Within minutes, the authorities review CCTV footage from the station, and are able to obtain a facial image of an individual leaving a suspicious package at the bomb-site minutes before detonation. The suspect is recognized by police, and is known to have links with terrorist organizations. The police know that the suspect will go into hiding, and if they do not find him immediately, apprehending him will be very difficult. The clock is ticking. The police maintain an integrated system of surveillance cameras at public, high-traffic areas throughout the city, all of which are connected to a central facial recognition database. A photograph of the suspect is enrolled in the system. Almost immediately, the system raises alarms from a series of cameras on the other side of town, allowing police to track the suspect's movement in real-time. Local police mobilize, and the suspect is apprehended 15 minutes after the detonation.

At the time of writing, wide-scale surveillance systems, as in the preceding scenario, are not technologically feasible. However, biometric systems have recently reached a level of performance where small-scale systems are possible, and larger systems may not be far off. Covert biometric surveillance is one of the most hyped, researched, and promising applications of biometric identification.

The primary focus of research into biometric surveillance is into hardware devices that can acquire a person's biometric from a distance, and the algorithms that conduct the match. However, in line with the theme of this book, our focus is not on the biometric technology itself, but rather system data and evaluation. Not only are surveillance systems on the cutting edge of what is possible for biometric identification, but they pose a number of significant challenges for evaluating accuracy. This is an emerging area, and has a number of complications as compared to the evaluation of traditional verification or identification systems. There are very few published studies or standards on the subject, and there are no evaluation competitions that directly involve a surveillance component. Therefore, much of the material in this chapter represents new, unpublished findings.

The most common and obvious biometric technology for surveillance is facial recognition. The advantage of this approach is that face matching algorithms have been widely researched, and highly specialized hardware is not necessary to acquire the biometric sample (i.e. a photograph of the face). Other biometrics that are under investigation for surveillance purposes include gait (the way people walk), iris (the research is focused on cameras that are able to capture high-quality images at a distance), ear shape, odor and a variety of others. All of these biometrics have their own significant challenges, so it is likely that the second generation of systems will be multimodal. In other words, they will fuse the results from a number of different biometrics, so the system is still effective when a particular biometric cannot be acquired or matched. This chapter is essentially a case study in face recognition surveillance. However, although face recognition is assumed, the same principles apply to any mode of biometric surveillance.

The following topics are covered:

- An overview of biometric surveillance, including a description of the problem and a discussion of the unique challenges (Sect. 11.1).
- Examination of issues related to the design and installation of facial recognition systems (Sect. 11.2).
- Discussion of techniques for running surveillance trials (Sect. 11.3) and evaluating the results (Sect. 11.4).

11.1 Biometric Surveillance

Covert biometric surveillance systems are designed to monitor an area and automatically identify persons of interest as they pass through. Without biometric technology, this is a very laborious job. For example, at the present time identifications for criminal investigations are predominantly manual, with human investigators poring through hour after hour of CCTV footage in the desperate hope of recognizing someone or something. The amount of footage reviewed is limited by the amount of man-power available to view the tapes. This is a natural application of pattern recognition technology. It is unlikely that the process will ever be fully automated, however at the moment the process is so time consuming, and the outcome so critical, that the use of biometric tools is inevitable.

Section 11.1.1 contains an overview of how surveillance systems are setup, and how they work. The unique challenges for this mode of identification are outlined in Sect. 11.1.2.

11.1.1 Surveillance Systems Overview

The goal of covert surveillance systems is to detect persons of interest without their knowledge or direct involvement. Figure 11.1 contains a schematic of the typical stages in the identification process. The stages are as follows:

Step 1   A person walks through an area under surveillance, known as the *capture area*. For a given camera, the physical area that it is has been designed to monitor is known as its *capture zone*. In the diagram, there are two hidden cameras, with two distinct capture zones.

Step 2   Each camera continuously monitors its capture zone and transmits the digital images to the back-end processing system. Some cameras may be fitted with motion detectors to reduce the storage and computational load of the data processing.

Step 3   An algorithm is applied to the input images to segment the biometric data from the background. In the case of face recognition, this algorithm is known as a *face finder*, and the output is sometimes referred to as a *token* (an image that only contains a face). Face finding algorithms are typically able to extract multiple faces from a single image, thereby supporting situations where groups of people pass through a capture zone simultaneously.

Step 4   The token images, or *probes*, are matched against each image belonging to the *watchlist*, also known as the *gallery*. The watchlist is the set of enrollments for people who are targets of the system. The matching must be conducted in real-time because if the individual walking through the capture zone actually does belong to the watchlist, this information must be known immediately so that appropriate action can be taken before they leave the area.

Step 5   If a probe matches a watchlist image with a score above a pre-determined threshold, an *alarm* is raised to alert the operator of a potential match. The alarms are presented to the operator in the form of a *candidate list*. A candidate list contains likely matches, which are reviewed by the operator. If an operator believes a match is genuine, it is labeled as a *hit*. An example of a candidate list can be found in Sect. 11.4.3, Fig. 11.2.

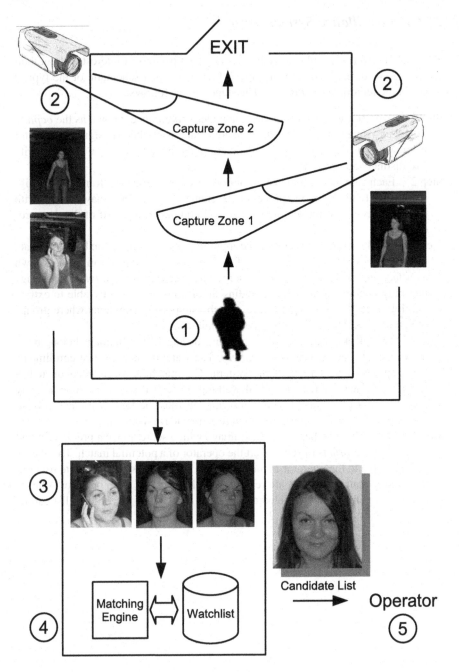

Fig. 11.1 Surveillance system overview. (1) A shadowy figure unknowingly passes through the capture area. (2) The hidden cameras record images from the capture zone. (3) The "face finder" algorithm detects and extracts faces from the surveillance images. (4) The facial images are matched against the watchlist by the matching engine. (5) A list of potential matches is returned to the operator.

Presenting the Candidate List

The candidate list is a very important aspect of the surveillance system. In many applications, the system as a whole can be viewed simply as a tool to reduce the workload for human operators who are monitoring an area. It accomplishes this by a) segmenting still face images from the background, and b) filtering out the majority people, who are unlikely to belong to the candidate list. Often the operator will be under heavy time pressure to make a decision as to whether or not a match is a correct hit. Therefore, the role of the user interface for surveillance systems plays a greater role than for other biometric applications. It is vital that the candidate lists are conveyed to the operator in an intuitive manner, with interactive image processing tools to aid image comparison.

11.1.2 Surveillance Systems Challenges

The process outlined in the previous section is conceptually straightforward. However, there are a number of difficulties unique to biometric surveillance that make it particularly challenging. First and foremost, covertly acquiring high quality biometric samples is difficult. This is primarily due to the numerous sources of variation that are difficult to control without direct intervention. The following is a list of factors that make it difficult to obtain high quality facial images:

- In many situations there will be considerable restraints on the site selected for the installation. In fact, site location may be a matter of necessity, and the system integrators have no choice but to work with what is available. Therefore, the environment at a site may be unfavorable for capturing high quality images:

  - Lighting is the primary environmental factor that impacts image quality for covert face recognition systems. An ideal photo for face recognition is uniformly illuminated from the front. If the lighting it too bright or too dark, the image contrast will be low, and fewer distinguishing features will be captured. Uneven lighting will lead to shadows, which can also hide distinguishing features. Another problem is back-lighting, where subjects are illuminated from behind. In this case, not only will the face be dark, but the light may spill over, and wash out a portion of the face. Under normal circumstances, proper lighting is achieved by using a flash, which is obviously not an option for covert applications. Furthermore, introducing bright lights to an environment may be conspicuous. Finally, the sun is a complicating factor for sites with a lot of natural light. In this case, there is an additional complication that the source of light moves throughout the day.

- The physical layout of the capture area will have an impact on image quality. For example, high ceilings may necessitate the placement of cameras far away from the capture area. In general, the arrangement of cameras is closely tied to detection rates. Therefore, poorly placed cameras will lead to inadequate coverage of the capture area, and result in high system error rates.
- Another environmental factor is possible occlusion due to pillars or other obstructions in the area.

- The role of behavior is amplified in surveillance systems due to a lack of direct interaction with the user. Therefore, behaviors that have minor effects in normal systems may be a considerable problem in a covert situation:

  - Neutral expressions are best for face recognition systems. However, for surveillance applications there is no control over the expression of the subject since the subject does not know that they are being photographed. Therefore, the full range of facial expression is possible, which can vary from neutral to smiling, frowning, yawning, talking, eyes closed etc. The authors have observed wildly contorted mid-sneeze captures.
  - Pose is the position of the face relative to the camera direction. Face recognition systems can generally handle a minor degree of tilt and rotation, but perform best when people look directly at a camera. However, for covert systems there is no way to enforce that the subjects to look directly at the cameras. This may result in people walking through the capture zone with their head facing up or down, or side to side. Furthermore, their gaze is free to wander around the environment.
  - The clothing people wear also has a particularly strong impact for surveillance. In a truly covert system there is no control over subject clothing. Hooded sweatshirts, scarves, sunglasses and brimmed hats are particularly troublesome, as they tend to obscure important regions of the face.
  - Another complicating factor is that the subjects are usually walking at the time of capture. The speed can vary from a leisurely stroll (e.g. coming back from a coffee break) to a frantic dash (e.g. late for the 2pm flight). The implications of this are obvious: capturing sharp, well-focused images of moving objects is much more difficult than stationary objects. This is further complicated by the fact that there may be poor lighting in the capture area.
  - Some behaviors can lead to partial or full occlusion of the face. For example, speaking on a mobile phone will lead to some blockage, and depending on the position of the camera, this may hide a significant portion of the face. Another common problem occurs when a group of people walk through at the same time. In this case people in the rear, especially if they are shorter than those in front of them, can be completely hidden.

The practical result of the preceding factors is that the failure to acquire rate is high, and when images are captured, the quality is unreliable. One of the recurring themes of this book is that biometric performance is fundamentally related to data quality. Therefore, poor quality captures will result in poor recognition performance, regardless of the underlying match engine.

There are other challenges for surveillance systems, apart from the difficulties associated with data capture. These include:

- Not only are the capture images difficult to obtain, but there may be significant difficulties obtaining high quality enrollment images. This is because, unlike many biometric systems, the watchlist members are not willing participants. For example, consider a system that includes all current members of the U.S. FBI's most wanted list.[1] These images are typically from a variety of sources, and often very poor quality. The fact of the matter is that the people who are usually of most interest as surveillance system targets are elusive by nature, and one cannot rely on access to recent, high quality photographs.

- In some cases targets are aware that they are being sought. When this is the case, they may be actively trying to disguise themselves. One way to accomplish this is to disguise their physical appearance using makeup, fake beards, cosmetic surgery, etc. Alternatively, they can exploit the behavioral factors listed above to their own advantage. For example, by wearing a hat, scarf and sunglasses, and walking quickly with their head facing down, a person can become a phantom, practically invisible to facial recognition systems.

- As mentioned above, there may be little choice as to the location where a system is installed. This can lead to environments that are not well suited to data capture. However, there are other, more practical considerations as well. In many locations access to electricity may be unreliable, there may be inadequate heating or air conditioning, and no access to the Internet. Unfortunately, this is typical of many sites where covert surveillance would be most useful.

- The hardware requirements of real-time surveillance are substantial. For high traffic areas, each camera generates a huge volume of high-resolution images. The face finding algorithm must be applied to each image, extracting and storing every face it finds. Next, the faces must be matched against the entire watchlist, and present the results to an operator. All of this must happen in a matter of seconds, so there are considerable processing and storage requirements.

With these considerations in mind, one can begin to appreciate the difficulty of the task. Covert surveillance is definitely the most ambitious application of biometric identification currently under investigation.

11.2 Site Design

Site selection and design is vital for minimizing the impact of the challenges outlined in the previous section. In most situations there will be considerable constraints upon site selection, and the final location may not have been selected based purely

---

[1] http://www.fbi.gov/wanted.htm

on its suitability for data capture. Once a site has been selected, there are two conflicting design criteria: maximizing the probability of a good capture, and minimizing the visibility of the system. Inevitably, there will be challenges that are unique to any particular environment. Therefore, there is no universal solution to most of the problems encountered during deployment. When these problems arise, a degree of creativity and ingenuity is an absolute necessity for a successful installation.

11.2.1 Site Selection

In most cases there will be limited choices available for suitable sites to select from. The following factors should be considered for site selection:

- The first and foremost consideration is selecting a location where the watchlist members are likely to pass through. Usually this depends on the nature of the application. For example, consider a casino that is trying to enforce a ban on known card cheats. In this case, cameras at the casino entries and exits, as well as above the card tables, is the natural choice. In other cases, there will be little a priori knowledge as to the whereabouts of the targets. In these cases, the best sites are high-traffic areas, such as train or bus stations. For applications related to immigration and customs, the suitable sites are restricted to entry and exit points to the country.
- For covert installations, another consideration is installation and maintenance access. Surveillance systems are difficult to install and tune, and some degree of private access is necessary.
- The goal of the capture zones is to select a series of small, well defined locations where people stop, or pass through, with high probability. These are known as *choke points*, and they will exist naturally at good sites. In general, the smaller the choke point, the easier one is able to focus the detection efforts on that spot. Long and thin sites tend to be preferable to wide, uncontrolled areas. Some examples of natural choke points are corridors, escalators, entry or exit doors, turnstiles, and metal detectors. The placement of obstacles, such as furniture or decorations, can be used to create an artificial choke point by restricting one's movement through an area.
- One must consider behaviors specific to the sites under consideration. For example, imagine a system that uses cameras that are installed outside a building, and are designed to capture people as they enter. In this case, on rainy days people may be carrying umbrellas, hiding their faces entirely.
- The site should lack visual obstructions. For example, pillars in the room may reduce capture opportunities.
- A site should have suitable places to hide cameras. On one hand, in a bare, stark room it will be difficult to inconspicuously place the cameras. On the other hand, a busy room full of clutter will have many possibilities.

- Lighting is a very important consideration. If a room is poorly lit, the introduction of new, bright lights may be suspicious. Therefore, wherever possible rooms that have a considerable amount of existing light should be favored.

11.2.2 Camera Placement

Camera placement and tuning is an established discipline in itself, and a detailed discussion of is beyond the scope of the book. However, from a biometric point of view there are a few general principles that should be kept in mind. The primary motivation should be to maximize the probability of a high quality capture for every person who passes through the area. Due to the uncontrolled nature of the system, everyone who passes through will behave in a slightly different manner. Therefore, it is generally not possible for a single camera to provide adequate coverage.

The previous section introduced the idea of a choke point, which is a small, well defined region where people are funneled through. In order to maximize the probability of a capture, it is usually necessary for several cameras to be focused on this area. The cameras should each give a unique perspective of the choke point. For example, subjects vary in height, and may be looking slightly to the left, right, up, or down. Therefore, the cameras should be arranged to give a number of different views, in the hopes that at least one will obtain a frontal image. The parameters for each camera that can be adjusted are height, direction, and focal length, all of which define a camera's capture region. This region is a conic section in 3D space.

The actual number of cameras necessary to adequately cover a capture zone depends on a number of factors. One factor is the frame rate of the cameras. If people at the site location walk quickly, it may be possible that a single camera will not have an opportunity to photograph everyone. In this case, multiple cameras may be necessary to compensate for a low frame rate. Another factor that must be considered is the degree of behavior variance. Some situations are more conducive to uniform behavior than others. For example, people tend to act in a similar manner as they step off an escalator. Specifically, there tends to be a brief moment as they step off where they look ahead, and contemplate where and when to step. On the other hand, there are fewer predictable behaviors when people are walking and talking in a crowd through an unconstrained environment. In these situations, more cameras will be necessary to capture images of sufficient quality.

In most situations, the only way to determine the optimal number of cameras is through trial and error. It is almost impossible to predict ahead of time all the factors that will influence subject behavior, let alone using this information to estimate the number of cameras necessary. As a general principle, more cameras increase the probability of high-quality captures, however there are practical limits to the number of cameras that can be installed. For one thing, at a certain stage there will be diminishing returns with the addition of new cameras focused on the same region. Also, there are serious hardware limitations that must be kept in mind. When crowds of people walk through a capture zone simultaneously, there are considerable real-time processing demands that increase with the addition of each new camera. Finally, adding new cameras increases the total number of captures, thereby increasing the frequency of false alarms.

It is often a good strategy to have more than one capture zone. This will help to avoid the problems associated with "putting all your eggs in one basket". There will always be events that lead to failed acquisitions. For example, consider a person who is completely blocked behind another person while walking through a capture zone. In this case, no number of extra cameras will guarantee a front-on capture. However, having a completely separate capture zone may give the system a second chance.

Having a plan for camera placement is important, but it is inevitable that the end result is not exactly as expected. Sites vary widely, so there are a number of aspects of site design that will need to be determined empirically through an iterative process of trial an error. Therefore, when planning a deployment, one must always be flexible, and allow sufficient time for adjustments after the initial installation.

11.2.3 Covertness

A covert system is one that is designed to blend into its environment. This poses several unique challenge for biometric surveillance systems. First of all, the cameras must be placed in such a manner that they are not clearly visible. If people know that they are under surveillance, they may alter their behavior to avoid looking in the direction of the cameras. This is especially true for people who are trying to hide their identity, which is likely for targets for such systems. Cameras can be hidden behind one-way mirrors, or placed in dark cavities where they are not visible from a distance. Signs, such as an illuminated exit sign or an advertisement, can often be introduced to a room when no suitable spot for hiding a camera already exists. It may also be possible to introduce disguised towers to house the cameras. For example, lamps or sculptures can be designed to hold several cameras, while blending in naturally with the environment.

The second challenge of covertness is influencing subject behavior. Since the systems are hidden, they cannot directly control the behavior of the people it is monitoring. However, there are some ways to impact a person's behavior without their conscious knowledge:

- In some environments a certain level of direct interaction is acceptable. For example, at an airport security staff can ask people take off glasses as they pass through a metal detector. Another example is security at a sporting event or music festival, where they can stop people and have them face a specific direction while being searched. In these situations, creative tricks can be used to increase the probability of frontal capture.
- When no human interaction is possible, *attractors* can be used to catch someone's attention, making them look in a desired direction. An example of an attractor is a brightly lit sign with a strongly worded warning. Attractors are especially useful

when they can house a camera, as then they serve a dual purpose. In general, bright colors and movement are known to stand out, and thereby draw attention.

- The opposite of attractors are *detractors* that make people look away from them. An example of a detractor might be an obvious, but fake, surveillance camera. Someone deliberately trying to evade detection would make a point of not looking in that direction.
- Obstacles can be useful in a capture zone for several reasons. First of all, they tend to slow people down, which increases capture opportunities. Secondly, they can be used to control the flow of people and create choke points. Finally, obstacles cause people to be more aware of their surrounding, lifting their heads to navigate. Once again, this increases the probability of frontal captures.

Lighting is an important factor for data capture. Ideally, a site will already have adequate lighting before installation. When this is not the case, additional lighting can be introduced by increasing the power of existing lights or adding decorative lamps. Sometimes it is possible to add a lighting component to attractors, such as installing brightly illuminated signs.

11.3 Running Scenario and Operational Evaluations

Recall from Chap. 5 that there are three types of biometric evaluations: technology, scenario, and operational. Technology evaluations compare algorithms using standardized test sets. In this case, the only uncontrolled variable for the experiments are the matching algorithms, allowing for a fair and direct comparison of error rates. It has been demonstrated in this chapter that biometric surveillance systems are particularly sensitive to the environmental and behavioral conditions in which they are installed. Therefore, a technology evaluation that focuses only on algorithmic differences between vendors is of limited interest. Although the performance of the matching engine is important, it is only one of several factors that determine performance, and in many situations it is not even a dominant factor. For example, if the lighting for a particular environment is leading to a high failure to acquire rate (i.e. many people are passing through the capture area with no pictures taken), the specifics of the matching algorithm are irrelevant. Therefore, in order to properly evaluate surveillance capabilities, one must model different environments, user behaviors, data capture systems, etc. These are taken into consideration by scenario and operational evaluations, which test systems as a whole.

With a scenario evaluation, the system is tested in an environment that simulates, as closely as practical, a real operating environment. For example, a site would be chosen that is representative of the types of places where a real system might be deployed. Volunteers are recruited, and told to walk through the capture area at specific times, testing for a variety of scenarios. If multiple vendors are being compared, each vendor will install their own data capture hardware.

If test subjects are aware that they are participating in a covert surveillance evaluation this may alter their behavior and introduce systematic biases into the results.

For example, a knowledgeable test subject may slow down and look for the hidden cameras if he or she knows they exist. This would artificially increase the probability of a high-quality capture. Therefore, when running a scenario evaluation participant knowledge should be restricted whenever possible.

An operational evaluation tests a system in an actual application environment, with a real target population. For example, the system could be installed in a public place, and observe people without their knowledge. In this case, there is no control over who passes through, when they do so, and how they behave. This represents an operational reality, so the detection and false alarm rates will bear the closest resemblance to expected performance rates.

Both scenario and operational evaluations have advantages and disadvantages. For scenario evaluations, there is control over the testing process, so creating watchlists, recording ground truth, and observing performance in real-time is relatively easy. However, the results are only relevant to the extent that true operating conditions can be predicted and modeled. This is very difficult, so there will always be a high degree of uncertainty in the observed rates. On the other hand, operational evaluations inherently model all the factors that impact system accuracy. The environment, target population and subject behavior are real. However, operational systems are also unpredictable, so creating watchlists and recording ground truth are difficult. The problem with creating watchlists is selecting people who will actually pass through the system during data collection. Recording ground truth is complicated by the fact that there is no direct intervention with the subjects, and if a target walks through undetected, this will generally not be known.

The most promising approach to surveillance trials is a combination of scenario and operational approaches. This involves using an operational environment with some minimal controls in place. For example, recruits may be randomly blended in with the real target population in a public place. Another example is using people at their place of work. In this situation participation can be voluntary, but the data capture system is set up covertly, so the targets do not know the specifics of the trial. When taking this approach, there are some ways to control the experiment without betraying the spirit of a covert installation. For example, a test can be set up to coincide with a fire drill, detecting people as they leave the building.

From the point of view of biometric data, the two most important aspects of running a surveillance evaluation are creating a watchlist and recording ground truth.

11.3.1 Creating a Watchlist

A watchlist is the set of enrollments for people who are targets of the system. In other words, an alarm should be raised when a watchlist member passes through the capture area. When creating watchlists for surveillance applications, the following factors must be considered:

- Enrollment image quality: Ideally the quality of the watchlist images should reflect the images that would be enrolled for a live system. In many cases, only poor quality images will be available for real persons of interest. If these are likely targets, a selection of poor quality images should be included in the watchlist.

- Template aging: The watchlist images should be captured well in advance of the evaluation. An image taken in the days or weeks before the test will normally not exhibit sufficient variation to model the effects of template aging. For a scenario evaluation, one strategy is to ask the participants to provide a personal photo taken months, or even years, before the trial. Note that using artificially altered images to account for template aging or poor data quality is not recommended.
- Demographics: The watchlist should reflect of the expected sex, ethnicity, and age range proportions of the target population.
- Watchlist selection: The watchlist should not contain every person who is expected to participate in the evaluation. In other words, it is important that some people walk through the capture zone who do not belong to the watchlist. As will be seen in the coming sections, this is important for false alarm analysis. Furthermore, there should be some people on the watchlist who do not pass through the capture zone during data collection at all. Both of these conditions are necessary to reflect real operational conditions.
- Watchlist size: It is well known that performance for identification systems is related to watchlist size (see Sect. 7.2.1.1). Basically, the larger the watchlist, the greater the probability of coincidental similarities, resulting in an increase in false alarms. However, some systems actually observe a performance degradation if watchlists are too small, due to the use of score normalization. Furthermore, the watchlist should contain enough people who actually walk through the system to ensure statistical significance of the detection rates (see Sect. 7.3.6). A rough rule of thumb is that a watchlist should contain about 300-500 individuals, about half of which should pass though the system. This number reflects many realistic operational scenarios, is small enough that the matching engine will not experience performance degradation and is large enough to allow statistical significance of the results.

The actual source of the watchlist images will vary. For a scenario evaluation, it is possible to collect a series of watchlists, testing for different degrees of quality and template aging. As mentioned above, the participants may be asked to provide a selection of personal images. Scanning passports or driver's licenses in an attractive option, as it reflects a realistic image source.

For an operational evaluation, the options for watchlist images are much more limited. If the targets are actual persons of interest, law enforcement agencies will generally have some photographs available. In other cases, it may be possible to collect watchlist images during the data collection itself. For example, consider a system that is installed near the immigration area of an airport. It may be possible to discretely collect passport images immediately before of after they pass through the capture area. In this case, the watchlists would be built after the completion of data collection. All capture images would be stored, and the matching would be run retrospectively to simulate the results.

11.3.2 Recording Ground Truth

Ground truth is a written record of what actually happened during data collection. It is comprised of three pieces of information: the identity of every watchlist member who passes through the system, when they pass through, and a synchronized time stamp for each capture image. Strictly speaking, it is not necessary to record the identity of non-watchlist members who pass through the system. However, this information may be useful for the analysis of the results. For example, it can be used to answers questions such as "are men or women more likely to cause false alarms?".

Ground truth is conceptually, yet deceptively, simple. Recording ground truth for a surveillance system is much more complicated than for conventional biometric systems, for a variety of reasons.

Firstly, in the section on evaluation it will be observed that surveillance events do not occur at specific instances in time, but rather within time windows when a person is "detectable" by the system. When groups of people walk through a system simultaneously, these windows overlap. Therefore, an ideal ground truth would record both the entry and exit time for watchlist members for each capture zone.

Secondly, with conventional biometric systems all relevant information is logged by the system itself. Events are triggered by user interaction, so logs are easily and automatically maintained. For example, consider an access control system. Every time a verification is conducted, there is active attempt at acquisition. Even if no image is captured, a "failure to acquire" event can be recorded. However, for surveillance systems the interaction with the system is passive. Therefore, if there is a failure to acquire (for example, due to a small person hidden behind a larger person), no image is captured, and the system has no record of the event. Therefore, one cannot rely on the system records; one needs to step outside, recording events and ground truth independently.

Finally, recording ground truth should be done with as little direct interaction as possible, as this may introduce behavioral biases into the results.

For a scenario evaluation, the most obvious approach is to explicitly record the times when people enter, and possibly exit, the capture area. This does not necessarily need to be manually intensive. For instance, it would be relatively straightforward and cost effective to use RFID to automatically log the time and identity of people as they pass through the capture zones.

For some environments, there may be existing systems that log identity and time. For instance, many buildings have areas that have swipe card access. A surveillance system could be designed to capture people as they enter or exit these areas, and have ground truth recorded with no additional effort. In this case, the watchlist would be comprised of a selection of people with access to the area. Also, records are usually maintained when people enter or leave a country. In this case, the watchlist can be comprised of frequent travelers.

For operational evaluations, recording ground truth during data collection may not always be practical. When this is the case, it is possible to establish some ground truth retrospectively. Non-empty candidate lists can be reviewed manually, and one can determine what proportion of alarms are correct detections, and how many are false alarms. However, this is an incomplete analysis as it *says nothing about missed detections*. In other words, it will be unknown how often a watchlist member passes through the system undetected. This is unfortunate, as this is usually the performance measure of most interest.

The Use of Video Footage

One common suggestion for recording ground truth is to use wide-angle CCTV cameras to give a bird's eye view of the scene. However, CCTV footage does not constitute ground truth. First of all, there is no guarantee that evaluators will be able to identify all the people from the video footage, which will suffer from the same environment and behavioral problems as the surveillance cameras. Secondly, this would be prohibitively manually intensive for all but the smallest evaluations. Therefore, CCTV footage may be useful for system diagnostic purposes, but should not be considered as a substitute for actual ground truth records.

Operator Logs

Another inappropriate method for storing ground truth is using operator logs. Operator logs are the result of a human operator examining the candidate lists, and deciding which matches appear to be correct, and which ones are false alarms. There are two fundamental problems with this approach. Firstly, in the event of an acquisition error or a low match score, the correct match will not be contained in the candidate list for an operator to verify. Therefore, information on missed detections is lacking. Secondly, operators themselves do not exhibit perfect recognition performance, and have their own associated error rates.

11.4 Performance Evaluation

After the trials have been run and ground truth recorded, the raw data must be crunched and converted into statistics that embody "system performance". By this point, it will probably not come as a surprise to the reader that this is more difficult for surveillance than for other biometric applications. This section describes the evaluation process. One simplifying assumption, discussed in Sect. 11.4.1, is that analysis is conducted for the system as a whole. Resolving ground truth is the process of using ground truth records to label actual capture images, and is presented in Sect. 11.4.2. Finally, methods for computing performance rates and graphs can be found in Sect. 11.4.3.

11.4.1 System Level Analysis

There are some tricky questions that must be addressed when planning an evaluation. Consider the following situation. A watchlist member passes through the system while wearing a ski mask. Obviously, a facial recognition system will be unable to identify him or her, in a trivial sense, because their face is not visible. In this case, should the system be penalized for failing to identify the person, when it really had no chance to begin with? Questions such as these lead to a large gray area. What if, instead of a face mask, they had been wearing a hat and sunglasses. In this case, a correct detection is almost as difficult. What if they have a habit of walking too quickly for a sharp capture? There are many examples of borderline cases, and any attempt to draw a line between cases that are "too hard for detection" and "should have been detected" is arbitrary.

The best resolution is obvious when one takes the focus away from the biometric matching algorithm, and remembers the goal of the surveillance system: to detect people who belong to the watchlist. As long as the reasons for a missed detection reflects an operational reality, a system should penalize missed detections, regardless of the underlying reason. However, it is important to stress that there are many potential causes for an error, and a *missed detection does not imply an algorithmic failure*. In fact, the ultimate cause for a missed detection will rarely be clear as disentangling the many potential factors is too complicated. It is likely due to a combination of factors, such as poorly placed cameras, camera optics (focal length, exposure times, etc.), time constraints on installation and tuning, and an inability of the feature extraction algorithm to account for expression, pose, or shadows. The key point is that it is not practical to control for all these variables, so the system as a whole must be evaluated. This includes the data capture system, environmental, and behavioral factors, watchlist quality, user demographics, etc. Biometric matching may be at the core, but it is certainly not the only factor influencing performance.

11.4.2 Resolving Ground Truth

As seen in Sect. 11.3.2, ground truth is comprised of three items of information: who walked through the system, when they did it and when the capture images were taken. However, this information in its raw form does not tell us what we need to know; namely, who the people are in each of the capture images. The process of labeling the capture images is known as resolving the ground truth.

In some cases there may be tens, or even hundreds, of thousands of unlabeled images. Therefore, automated techniques should be used whenever possible. However, the labeling process is inherently fuzzy, so some manual interaction will always be necessary.

11.4.2.1 Image Labeling

In the context of surveillance systems, define an *event* as a single person passing through the system. A *watchlist event* is an event for a person who belongs to the watchlist. In theory, a watchlist event should raise an *alarm*. For each event, there will be zero or more capture images. If no images are captured, it is known as a *failure to acquire*. On the other hand, in the situation that a person is temporarily stationary in a capture zone, dozens, or even hundreds, of images may be captured. Therefore, the number of images corresponding to a particular event can vary widely. Unlike most biometric systems, there is not a simple correspondence between identification transactions and the number of acquisition images.

There are two steps for labeling images:

1. The first step is to partition the capture images into two sets: those corresponding to watchlist events, and those not corresponding to watchlist events. The first set represents potential detection opportunities, and the second set may lead to false alarms. This is done as follows. Each capture image has a time-stamp of when the photograph was taken. This time can be cross-referenced with the ground truth in order to determine if any watchlist members passed through the area at this time (i.e. the time-stamp for the capture image is within a watchlist event's time window). Depending on the frequency of watchlist events, this step can automatically label a significant portion of the capture images as not containing watchlist members.

2. The two sets from the first stage need to be further refined into: capture images containing labeled watchlist identities, and capture images not containing watchlist members (in this case, the actual identity is not important). It is already known that images in the second set above do not contain watchlist members, as they do not correspond to watchlist events. Therefore, these images do not need to be revisited. However, it cannot be assumed that images from the first set contain watchlist members. For example, there may be several watchlist members walking through at the same time, and there is a need to determine which image belongs to whom. Also, a false alarm may occur at the same time as a watchlist event, and this needs to be verified manually. However, the task is not as laborious as it seems at first (assuming the availability of proper tools), as the number of watchlist members passing through the system at the time the image was captured will usually be one, otherwise limited to a handful. One need only confirm which, if any, of the small number of a watchlist identities a capture image corresponds to.

The amount of manual effort necessary for image labeling depends on the precision of the ground truth data. If the event windows are tight, such as a few seconds, most non-watchlist captures will fall outside of watchlist event windows. Furthermore, most watchlist captures will fall within the corresponding watchlist event's time window, and so labeling can be largely automated. This reinforces the importance of gathering accurate ground truth data. The effort invested in properly recording ground truth during the trials will pay off greatly during the analysis.

If the ground truth is not accurate, it may be the case that there is a large degree of overlap between watchlist event time windows and non-watchlist captures. In other words, almost every capture could potentially belong to a watchlist member. When this is the case, it is tempting to use a match engine to aid with the resolution process. If done properly, this may be of some use. However, in general it is not recommended as there is a fundamental problem with evaluating face recognition performance based on the results generated from the same, or even a different, face recognition engine. However, using the match score results is a useful, if not necessary, post-processing step.

11.4.2.2 Post-Processing

The previous section presented a two-step approach to labeling the capture images. In the first step, the images were partitioned based on whether or not they corresponded to a time when a watchlist member was passing through the system. In other words, the capture images were divided into sets "may contain a watchlist member" and "does not contain a watchlist member". The second step used visual inspection to remove false matches from the first set, and label the identities of the remaining images. This is an inherently uncertain process, and it is almost impossible to avoid making some errors.

There is one further piece of information available that can be used to aid the resolution process. Recall that during data collection each image captured is compared against the entire watchlist and assigned a similarity score. Until now, a conscious effort has been made to avoid the use of these scores while resolving ground truth. However, they can now be used to a) fix obvious errors, and b) gain confidence in the ground truth resolution. This primarily involves examining outliers. For example, consider a labeling error that assigned genuine status to an impostor match. In this case, the capture image and watchlist image contain different people, and therefore the match will likely receive a low score. Therefore, these cases can often be detected by examining low scoring matches labeled as "genuine". On the other hand, consider a genuine match that is mistakenly put in the "does not contain a watchlist member" category. In this case, the match would tend to get a high score, and manifest itself as an impostor outlier. Looking at these outliers allows an opportunity to perform data cleansing, and will provide a sense for the accuracy of the automated labeling procedure.

11.4.3 Performance Measures

The goal of the surveillance system is to return a candidate list to the operator, which is a list of likely matches. With surveillance systems, it is not known a priori if the probe belongs to the watchlist. This is known as open-set identification, and is discussed in Sect. 7.2.2.2. There are two approaches to candidate list selection:

Alarm threshold   When an alarm threshold is defined, all matches that receive a score above it raise an alarm, which is an indication that a potential match has been found. Alarm thresholds are used when it is only deemed necessary to alert the operator when a likely match has been found, and not present candidates for every person who passes through the system. The alarm threshold is an adjustable parameter of the system, and is determined by balancing the false match rate against the detection rate.

Rank   In some cases, an operator may wish to observe the top ranked matches for every person who walks through the capture zone. For example, the top 5 matches from the watchlist may always be displayed. In general, this is much more laborious for an operator, and is only a feasible approach when a) the volume of people passing through the capture zone is low, and b) it is known in advance that there is a high probability that people belong to the watchlist.

When a person walks through a capture zone, there are four possible outcomes, depending on whether or not they belong to the watchlist. Each possibility is illustrated in Fig. 11.2. In the case that the probe contains a watchlist member, there can be a:

Correct detection   A correct detection occurs when a watchlist member belongs to a candidate list. When an alarm threshold is in use, this means that they have matched their own watchlist image with a score above the alarm threshold. This is illustrated in the first row of Fig. 11.2.

Missed detection   A missed detection occurs when a person belongs to the watchlist, passes through the system, but fails to be included in a candidate list. There are two possible causes for this. The first is that the capture system failed to obtain an image of the subject. In this case, there would be no image for identification. The second possible cause is that an image was captured, but it scored (or ranked) too low to be included on the candidate list. In the fourth row of Fig. 11.2, the probe achieves a score of 98 for the correct match. Assuming an alarm threshold of 100, this would not be returned as part of the candidate list, so is not considered a correct detection. However, it is the highest ranked match, so for a rank-based system, the individual would have been identified.

When a person walks through the capture area who does not belong to the watchlist, there are two possibilities:

Correct non-detection   This occurs when an empty candidate list is returned, and is only relevant for systems based on an alarm threshold. This is illustrated in

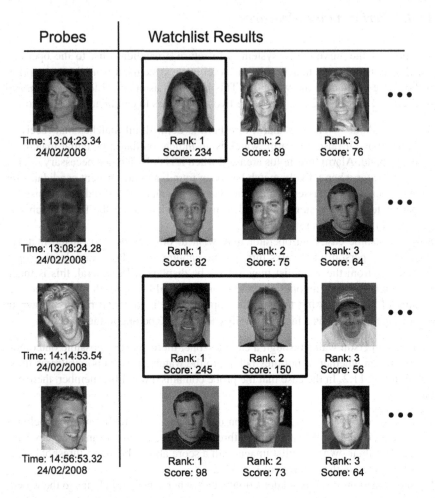

Probes | Watchlist Results

Time: 13:04:23.34
24/02/2008

Rank: 1
Score: 234

Rank: 2
Score: 89

Rank: 3
Score: 76

Time: 13:08:24.28
24/02/2008

Rank: 1
Score: 82

Rank: 2
Score: 75

Rank: 3
Score: 64

Time: 14:14:53.54
24/02/2008

Rank: 1
Score: 245

Rank: 2
Score: 150

Rank: 3
Score: 56

Time: 14:56:53.32
24/02/2008

Rank: 1
Score: 98

Rank: 2
Score: 73

Rank: 3
Score: 64

Fig. 11.2 Sample surveillance results. The probe column contains images covertly obtained from the data capture system. The watchlist results contains the top 3 ranked results from the watchlist, along with the score. The watchlist results in a box are the candidate list if an alarm threshold of 100 is in use. The first row is a correct detection, the second row is a correct non-identification, the third row is a false alarm and the fourth row is a missed detection.

the second row of Fig. 11.2. In this case, no matches have scored above an alarm threshold of 100, and the probe does not exist in the gallery.

False alarm   When an alarm is raised for an impostor match, it is known as a false alarm. For example, in the third row of Fig. 11.2, a two item candidate list has been returned, however neither of the people in the candidate list are a correct match with the probe.

There is some ambiguity in the definitions given above. For example, consider the case where a candidate list is returned that contains both a correct match and an incorrect match. In this case, is it a correct detection, a false alarm, or both? This question was previously encountered in the section on open-set evaluation (Sect. 7.2.2.4). There is no definitive answer, and it depends to a large degree on system policy. In most cases, it will be sufficient to consider this as a correct detection, which is intuitively satisfying, and helps simplify the computation of the error rates.

Another complication is related to the fact that multiple images of a person may be captured as they pass through the capture area. This is likely because the cameras often have a high frame rate (e.g. 20 frames per second), and there may be multiple cameras and capture zones. Imagine a person has four images captured, all of which matched above the alarm threshold. If they belong to the watchlist should this be considered four distinct correct detections? What if two instances are correctly detected, but two are not? On the other hand, if they do not belong to the watchlist, should four high matching images of the same person be considered four false alarms?

These answers are resolved in the following sections, and the key idea is to treat detection rates and false alarm rates independently.

11.4.3.1 Detection Rate Analysis

In general, it is reasonable to assume that a watchlist member is either detected or not detected, but cannot be detected multiple times. Therefore, detection rate analysis is based on watchlist events, not on capture images. For each watchlist event, there are zero or more capture images. If any of these images achieve a score greater than the alarm threshold, the person has been detected. The following are the high-level steps for computing the detection rates:

1. Gather the relevant data: One of the outputs of the image labeling stage (Sect. 11.4.2.1) was a set of labeled capture images that contain watchlist members. Along with the ground truth, this is the data necessary for the analysis.
2. Tabulate the genuine scores: Recall that a watchlist event is a time period when a watchlist member passes through the system. For each watchlist event:

   a. Select the zero or more capture images that correspond to this event. In other words, collect all the images of this person that were captured for this event.
   b. Since this person is a watchlist member, each image will have a genuine match score that was computed when the capture image was matched against the watchlist. Select the highest score achieved for this event. If there are no capture images corresponding to this event (i.e. a failure to acquire), assign this event a score less than the minimum possible score for the system. For example, if the score range is [0, 1]), assign failure to acquire events a score of -1.

3. Compute the detection rate: There is now a single genuine score for each event, even those with no corresponding images. The detection rate is computed based

on these scores. At a given alarm threshold t, the detection rate is the proportion of event scores $\geq t$.

Note that failure to acquire events have been assigned a score outside the normal score range. The effect of this is that even at the minimum possible alarm threshold, 100% detection rates still may not be achievable. This accounts for cases where subjects pass through the system without any images being captured. When this occurs, no images are displayed to the operator, so detection is theoretically impossible.

11.4.3.2 False Alarm Rate Analysis

Detection rate analysis was conducted at the level of events. However, false alarm rate analysis is conducted at the capture image level. Therefore, in theory, a single impostor can raise multiple false alarms. The rationale for this is that it may be necessary for a human operator to examine every false alarm individually. Therefore, the false alarm rate is proportional to the effort required of the operator.[2]

The false alarm rate is computed as follows:

1. Gather the relevant data: During image labeling, a set of images was identified that did not contain watchlist individuals. Each of these images has the potential to cause a false alarm, and they therefore form the basis of the analysis.
2. Tabulate the rank 1 impostor scores: For each probe, select the highest score it received when matched against the watchlist. Since this person is known not to belong to the watchlist, this rank 1 score is guaranteed to be an impostor match.
3. Compute the false alarm rate: The false alarm rate is based on the highest impostor scores for each probe. At a given alarm threshold t, it is the number of rank 1 impostor scores $\geq t$.

11.4.3.3 Computing the Alarm Graph

The computation of the alarm graph is based on the rank 1 genuine distribution and the rank 1 impostor distribution. Alarm graphs were covered in detail in Sect. 7.2.2.4.

Figure 11.3 (a) contains the rank 1 distributions for a simulated system with an ERR of 2.3%, and a watchlist of size 500. Note that there is a bump in the genuine distribution at -5, which is due to failure to acquire events. The alarm graph for the data can be found in Fig. 11.3 (b). For example, consider the point at a false alarm rate of 1%. This means that approximately 1% of non-watchlist members who pass through the system will raise a false alarm. At this operating point, 40% of watchlist members who pass through the system will be detected. Note that even at a false

---

[2] Note that some systems will automatically combine multiple images of the same person, and only display a single image to the operator. This is useful for reducing multiple false alarms due to the same person. However, as this happens "under the hood", it is not a factor that needs to be considered during analysis.

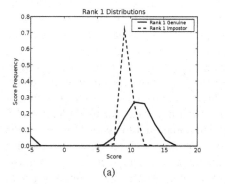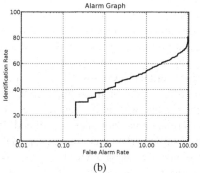

Fig. 11.3 Surveillance alarm graphs. (a) The rank 1 distributions for the genuine and impostor match scores. (b) The corresponding alarm graph.

alarm rate of 100% (i.e. all capture images raise an alarm), the detection rate is only 80%. This is due to failures to acquire, which essentially bound the maximum possible detection level.

When observing false alarm graphs for surveillance data, it is necessary to keep in mind the distinction between what is measured on the y-axis (the detection rate) and what is measured on the x-axis (the false alarm rate). The detection rate is based on events, each of which may have any number of associated capture images. On the other hand, the false alarm rate has a one-to-one correspondence with capture images. Despite this difference, the relationship between the two still represents the ubiquitous biometric trade-off between false accepts and false rejects. The alarm threshold must be adjusted to balance these rates, and an improvement in one is always at the expense of the other.

Alarm graphs are important because they present the system over a range of thresholds. Often, people unfamiliar with the evaluation of biometric data think that a match just beneath the alarm threshold is "good enough", and so should be considered a match. However, one must always keep in mind that it is necessary to fix a particular threshold, and by lowering this threshold one will increase the incidence of false alarms.

11.5 Conclusion

At the time of writing, biometric technology is reaching a level of accuracy where it is on the verge of becoming feasible for real-world, covert surveillance systems. It is certainly one of the most hyped, and controversial, applications of biometrics. There are a number of obstacles that make this a difficult problem, most of which are related to covertly sampling a person's biometric from a distance. This is a significant technological challenge, and an enormous amount of research is being invested in the problem. However, it is important that researchers focus not only on the sensing and matching technologies, but also on accurate methods for testing and evaluation. If this area is neglected, there will be inevitable disappointment when systems are put into production, and vendors fail to live up to their expectations.

This chapter has not addressed all of the issues related to conducting and evaluating surveillance systems. However, it is the most extensive treatment of the subject published, and is an important step in the direction of formalizing the techniques for surveillance evaluation. Several areas have been neglected; for example, operator performance has not been mentioned. It is important not to assume that operators will exhibit perfect recognition, especially in real-time, high-pressure environments. Operator performance will be related to a number of factors, such as their familiarity with the watchlist subjects, and the quality of the watch and capture images. In some situations, operator performance may actually be worse than that of the matching engine. Therefore, it is recommended that operator trials are run in parallel with evaluating the biometric component of a surveillance system.

Another important area that has been neglected in this chapter is the social and ethical implications of covert biometric surveillance systems. This is currently the matter of much discussion and debate. On one side, there are those who claim that surveillance systems will increase public safety and security. Terrorists and criminals will become increasingly sophisticated, exploiting all new technologies at their disposal. Therefore, every means at our disposal for fighting crime should also be used. On the other side of the debate are those who feel that the technology infringes on personal privacy [1]. They fear an Orwellian future where ubiquitous surveillance systems silently, and indiscriminately, monitor the public. One thing is clear - as the systems become more common, public perception will play an important role in the adoption of the technology. Therefore, one important step is that legislation is established to regulate the use of the technology. Furthermore, advocates should engage and educate the public while the field is still young.

References

[1] Agre, P.E.: Your face is not a bar code: Arguments against automatic face recognition in public places. http://polaris.gseis.ucla.edu/pagre/bar-code.html (2003)

Chapter 12
Vulnerabilities

The assessment of vulnerability is vital for ensuring biometric security, and is a concept distinct from system accuracy. A perfectly accurate biometric system may still be highly vulnerable to attack, as unauthorized users may find alternates ways by which they can be falsely accepted by a system.

Compared with the effort expended on determining performance accuracy, significantly less effort has been given to the problem of determining if a presented biometric is real or fake. With the increasing use of biometric systems, the understanding of vulnerability related risks and their appropriate treatment will be a vital part of future biometric deployments.

All the attack methods described in this chapter are vulnerabilities that are publicly known. As a general principle, the public dissemination of points of vulnerably is an important step towards ensuring system designers can put in place appropriate risk mitigations. Secrecy about avenues of attack can help potential fraudsters more than the disclosure of risks, since where the risks are not understood by the system owners, attack methods may be easily exploited. The principle of security through transparency is accepted practice in the cryptographic community.

There are four high-level factors that contribute to a biometric system's vulnerability to a determined attack: the security of the computing infrastructure, trust in the human operators, the accuracy of the matcher, and the ability to detect fake biometrics. There are long established standards and practices for assessing the first two factors, which include ensuring the security of communication channels and storage, tamper-proofing devices, and establishing usage policies. However, the non-deterministic nature of biometric matching and its interface to the real world means the latter factors create a variety of new security threats to be mitigated against in order to reduce the chance of a malicious attack being successful.

The goals of this chapter are to:

- Introduce the analysis of biometric vulnerabilities (Sect. 12.1).
- Look at the research that has been undertaken in detecting fake biometrics (Sect. 12.2.2).
- Outline the different points of attack in a biometric system (Sect.12.3).

- Describe different fraud types, including enrollment, covert and cooperative (Sect.12.4.1).
- Discuss methods for assessing vulnerabilities (Sect.12.5).
- List mitigations that can be used to address biometric vulnerabilities (Sect.12.6).

Definitions for terms related to vulnerability used in this chapter can be found in Chap. 6.

12.1 Introduction

In February 2002, former U.S. Secretary of Defense Donald Rumsfeld said in response to questions about military threats:

> "... there are known knowns; there are things we know we know. We also know there are known unknowns; that is to say we know there are some things we do not know. But there are also unknown unknowns – the ones we don't know we don't know." [5]

The detection and mitigation of biometric threats involves many of the same questions about what is known and unknown. The list of "known threats" is those that can be, or already have been, identified. Each "known threat" that has not been able to be investigated or evaluated fully can then be categorized as a one of the "known unknowns". In other words, the threat is known, but its impact or likelihood is not well understood. Threats may also be "yet to be discovered", and these are the "unknown unknowns". The discovery of these new threats requires active intelligence on the activities of attackers and researchers. It also requires creative and knowledgeable vulnerability evaluators. One thing that seems certain is further attack methods will be discovered, so the quick identification and mitigation of vulnerabilities is increasingly important to both the security of systems and the credibility of the industry.

The cost of any particular vulnerability is proportional to the value of the assets protected by the biometric, which might range from secure access to an office to the launch control for a missile, multiplied by the total risk of compromise, which is the chance that an attack will be successful. It is the total factor risk or the "spoofing risk" that is often of interest when examining a system's vulnerability.

Biometrics is a probabilistic science. Every time an individual has their biometric acquired it will be slightly different. This variation is caused by a combination of user behavior, environmental conditions and physical aging of the biometric, and means we can never be absolutely certain of identity through biometric means alone. However, the vulnerability of a biometric system should not be confused with its accuracy. It is possible to have a system that is extremely accurate at distinguishing between any two individuals, but which may be highly vulnerable to simple methods used to circumvent the security, either by mimicking physical biological characteristics, or by bypassing or altering the information used as part of the matching process.

In order to develop commercially useful biometric systems, past effort has focused on improving the ability of the biometric algorithms to distinguish between different people. Most large-scale evaluations of biometric technology conduct tests to determine the probability that a random person will match successfully against them self, or be mistaken for someone else. As a consequence, matching engines have become increasingly specialized in undertaking this distinguishing task, and deliberately ignoring transitory factors that do not aid in the identification process. However, this focus can potentially increase the system's vulnerability to attack by reducing the number of aspects of a biometric presentation that an attacker needs to fake. Therefore, vulnerability mitigations should focus on techniques that are supplementary, or orthogonal, to improving algorithm performance. These techniques are often termed 'liveness' or 'spoof' detection. The goal of the liveness detection is to prevent the acceptance, regardless of the match score, of fake biometrics (Fig. 12.1).

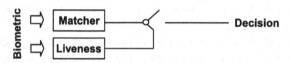

Fig. 12.1 The relationship between liveness detection, biometric matching and the match decision.

12.2 History

A variety of biometric vulnerabilities have been exposed both formally through academic research [13, 20, 27] and informally through magazine articles [26], hacker groups [15] and even television shows [6]. However, many of the demonstrated exploits are undertaken on less advanced systems, and examinations of vulnerability have been largely ad-hoc rather than systematic searches for all threat vectors.

12.2.1 Common Criteria

The international standard used for computer security, particularly by governments, is called the Common Criteria (CC) [2, 4]. The goal is to define rigorous standard processes that will determine a level of assurance, known as an Evaluation Assurance Level (EAL), in the security of computer products. For each security technology the requirements to be assessed are listed in a document called a Protection Profile (PP). Several such PP's are available from national standards bodies [1, 3, 14], however the most influential is the United Kingdom PP [25].

Unfortunately, the Common Criteria has not been very successful for the evaluation of biometric devices, as these are rapidly evolving technologies and have not been compatible with the sometimes ponderous and costly Common Criteria. The non-deterministic nature of the technology makes it harder to undertake formal testing, as this relies on strict repeatability and restrictions in the set of valid input parameters. However, this approach is not suitable for exploring the space of potential vulnerabilities for a biometric system.

12.2.2 Liveness Research

The detection of liveness is an active area of research, and for each biometric modality different techniques have been suggested for assessing liveness. Fingerprints have had the largest amount of research undertaken; suggested techniques include optical properties [18], pulse [23], perspiration [22], electric resistance [23], subepidermis structure [23], skin deformation [18], papillary lines[11], pores [18] or a combination thereof [18, 24].

For face recognition, the incorporation of head motion has long been known as a method to prevent the use of a static pictures [26]. Furthermore, the natural blinking rate of eyes can be used [21], as well as multi-spectral imaging. Face recognition systems can be particularly vulnerable to poor quality image enrollment [17], so ensuring quality control assists in preventing certain attack mechanisms.

The detection of pupil movement and saccade (eye movement) [9, 10] is used in some iris systems. Other techniques include the use of controlled light to check pupil response, the detection of infrared reflections off the cornea [9], and multi-spectral sensing [8] .

12.3 Points of Attack

A biometric system is composed of a number of different subsystems (see Sect. 1.6). Each subsystem may have a number of different points of attack, and for each point of attack there may be one or more potential exploits. Although such attack points exist in all matching systems, not all are equally vulnerable. In general, the less distributed the system, the easier it is to secure.

For instance, consider a secure biometric smart-card with all of the subsystems, including the sensor, integrated on the card. In the case that only the authentication decision need to be securely transmitted, this would significantly reduce the potential points of attack. Such cards may soon be practical with advances in manufacturing and sensing technologies. The other major factor is the degree of trust in the network and environment, as this affects the likelihood of a vulnerability being exploited.

The points of attack are shown in Fig 12.2. This diagram shows a biometric system both at the subsystem level as well as each individual component, and highlights the potential points of vulnerability:

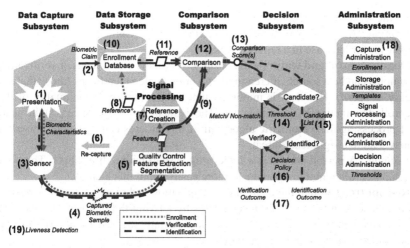

Fig. 12.2 Vulnerability points in a general biometric system. Derived from a diagram used with permission from Tony Mansfield, National Physical Laboratory, UK.

1. **Presentation**: The use of a fake biometric to enroll is the principal threat at the presentation point. This may be through the use of an *artifact* instead of the live biometric, or through the modification of an existing characteristic made to look more like an enrolled one, for instance the use of makeup. This process may occur at either the enrollment or verification stage. Another vulnerability is that a user might be coerced into presenting a biometric.
   Example: A latent fingerprint obtained from a glass surface and a fake fingerprint made from gelatin that replicates its ridge pattern is created and used to fool a fingerprint sensor.
2. **Identity Claim**: At the point of enrollment or verification, the use of a fake or stolen identifier to create a false claim of identity may result in either an invalid enrollment or a potential false accept.
   Example: An attacker creates a fake passport and uses this proof of identity to create a new identity in a government application that uses iris recognition.
3. **Sensor**: The integrity of the sensor allows trust in the integrity of the acquired biometric sample. If the sensor can be faked or compromised, the system may see what it believes to be a valid biometric sample, but which is actually an artifact or replay attack. Ideally the signal processing subsystem will be able to check using cryptographic techniques[1] that the sensor has not been tampered with and is operating properly.
   Example: An attacker removes the camera from a laptop and replaces it with a fake that always sends the same image.
4. **Transmission - Sample**: If the transmission of the biometric sample from the sensor to the signal processor travels over an insecure connection it may be intercepted for later use. Alternatively, the substitution of a fake biometric sample

---

[1] In smart-cards this is known as a SAM (Security Authentication Module)

for the real one may be undertaken.

Example: An attacker intercepts a fingerprint image coming from a sensor and stores it for later use in a replay attack.

5. **Quality control and feature extraction**: Low quality enrollments or verification samples can be a source of creating lamb or chameleon templates (those that are easy to spoof). Hence, it is important that tight control over the quality of the enrolled biometric data is maintained. The extraction of the features is at the heart of processing for a biometric algorithm: if this process can be compromised, it may lead to a significant threat.

Example: A person enrolls in a fingerprint system with a dirty finger. The poor quality enrollment allows others to more easily spoof this identity.

6. **Re-capture**: Through the continual re-acquisition and re-capture of a biometric an attacker can refine attack mechanisms by altering the biometric to discover which techniques work best. This may be especially the case for algorithms that are widely available for general purchase.

Example: A hand geometry system allows an attacker to attempt unlimited retries. This allows the attacker to figure out how to spoof the sensor.

7. **Reference creation**: If the creation of the reference feature and generation of a template can be compromised, this then can create a significant threat.

Example: A hacker inserts or changes code in the reference creation to ensure that whenever a particular palm vein pattern is seen it always generates a high score. The hacker can then distribute copies of this reference template to allow system access for fellow hackers.

8. **Transmission - Reference to enrollment**: If the transmission of the template from the reference creation process is over an insecure channel, the enrollment might be substituted for another before it is stored in the database.

Example: A hacker has infiltrated the database connectivity layer and substitutes templates as they are inserted into the database.

9. **Transmission - Features to database**: As for the enrollment reference transmission, when the features from the reference creation process are sent over an untrusted channel, the sample might be substituted for another before it is stored in the database.

Example: A hacker has infiltrated the database connectivity layer and substitutes templates before they are matched.

10. **Enrollment database**: The enrollment database is the source of the authentication data; if the enrollment database is compromised, this would allow any number of potential alterations and substitutions.

Example: A malicious database administrator inserts new templates for attackers.

11. **Transmission - Reference from database**: When the reference is transmitted from the database if it is over an insecure channel a hacker would be able to substitute templates before they are compared.

Example: A hacker substitutes a template retrieved from the database before it can be compared.

12. **Comparison process**: The comparison process creates a similarity score between the reference template and the verification sample features. If hackers can

compromise this process, they can output high scores for selected identities.
Example: A hacker changes the comparison process so that high scores are always given during a specific time period.

13. **Transmission - Score**: If the score is not transmitted securely it may be altered before it reaches the decision subsystem.
    Example: A hacker substitutes high scores for people with particular identities.

14. **Threshold process**: If the threshold process is compromised, the match threshold may be lowered, making it easier for attackers to be accepted by the system.
    Example: A hacker sets the system threshold to zero, allowing all individuals to pass.

15. **Candidate list**: During identification the candidate list results could be modified or re-ranked to exclude specific individuals.
    Example: A hacker ensures that particular identities are never ranked highly enough to be presented to the operator.

16. **Decision policy**: The decision policy uses business rules to convert the match results into a final acceptance or rejection. An attack who had the ability to change this policy would be able allow acceptance decisions at will.
    Example: Modification by a rogue administrator of the business rules around exception cases, falsely labeling an individual as someone who could not enroll, in order to bypass the biometric security mechanisms.

17. **Transmission - Outcome**: The final decision needs to be transmitted for action; if this transmission protocol is compromised then the matching outcome could be altered to generate a successful match.
    Example: A fingerprint sensor used for access control on a secure door transmits the unlock code to the door lock using a simple power relay. The attacker removes the sensor from the wall and shorts the open wires together, causing the door to unlock.

18. **Administration**: The administration subsystem potentially controls all aspects of decisions from acquisition and quality setting through to business rules and threshold settings. The security and audit of administrative access is hence a critical component.
    Example: A malicious administrator substitutes a fake enrollment and reduces the thresholds to allow an attacker to pass under a false identity.

19. **Liveness detection**: Liveness detection is the mitigation strategy used to protect against the use of prosthetic artifacts. However, the liveness detection process itself may also be open to attack.
    Example: A fingerprint system that uses heat for liveness detection may be spoofed by warming the artifact fingerprint or by creating a thin film to put over the top of a real finger.

12.4 Fraud

In addition to the attack points described above, fraud in biometric systems can be broken into a number of different classes. These dependend on when the fraud occurs (enrollment or verification) and the type of attack that has been mounted (covert or cooperative). Covert verification fraud is the most commonly considered fraud type, however it is the processes around enrollment that are the most crucial to ensure system integrity.

12.4.1 Enrollment Fraud

One of the most vulnerable points in any biometric system is the enrollment process. If poor control is maintained over the enrollment process then the overall integrity of the system can be seriously compromised.

Ensuring the enrollment processes has integrity usually means providing a reliable link to other identity credentials. This credentialing is commonly achieved through the use of proof of identity documents such as a birth certificate, passport or driver's license. *The strength of subsequent authentications using a biometric is dependent on the integrity and strength of the enrollment process.*

Where it is important to have high credential strength, biometric enrollment should always be supervised, as this helps mitigate against the use of artifact attacks since it is harder to use a fake biometric when you are being watched, and a human can also look for other suspicious activity. Also important is maintaining a strong audit trail of the enrollment process, including who undertook the enrollment and when it occurred.

12.4.2 Covert Fraud

Covert fraud is when an attack is undertaken without the knowledge of the person to be spoofed. This is the most common scenario to educate people about since it can lead to identity theft.

An attacker can covertly obtain an individual's biometric through several mechanisms. The most commonly considered is the creation of an artifact by use of either a biometric impression, for instance dusting for a fingerprint left on a surface and then creating an artifact, or through some form of surveillance activity. Often the covertly acquired biometric will be degraded through noise or missing features, and the creation of the artifact for attack will be of lower quality or have significant quality variations. The creation of artifacts will seldom have a perfect success rate, so any attacker needs to consider the risk that the artifact produced may in fact fail when tried on the target systems.

Raising Latent Fingerprint Prints

The fingerprints of the last person to use a sensor are sometimes visible on the surface of the sensor for some time. In early implementations of fingerprint sensors it was discovered that these 'latent prints' could be raised through simply breathing on the sensor or using a bag filled with water [26]. Modern systems should not be susceptible to such attacks since detecting the presence of a fingerprint that is almost identical to the one used previously is relatively easy. However, not all sensors explicitly check for this vulnerabillity.

Other methods of covert fraud involve reverse engineering the template. Since the template contains enough information for the algorithm to recognize a person, it follows that it should be possible to reconstruct a biometric that would successfully pass using only this information. Two methods have been used for this purpose. The first is called a *hill-climbing attack*. In this attack, the attacker must have access to the output of matching algorithm. They then use this to compare the template to some sample input. Random successive changes are made to the sample, and those that improve the match score are maintained and iteratively modified. Eventually, this may result in an fake biometric sample that doesn't necessarily look like the original, but is able to successfully authenticate. Examples using this technique have been shown to be effective in defeating face recognition systems [7]. The other form of attack is called a *masquerade attack*. This attack utilizes knowledge about the structure of the template (for instance, the position and location of fingerprint minutiae) to attempt to explicitly recreate the source biometric. It relies on the attacker understanding how to decode and interpret the template structure. Fingerprint systems have been successfully spoofed using this technique [16].

12.4.3 Cooperative Fraud

When the party to be imitated is colluding in the attack, for instance by allowing other people to use their identity, it is considerably easier for the attacker. Cooperative attacks might include the deliberate creation of poor quality templates (lambs), or the use of an artifact during enrollment that can be given to others.

'Insider fraud' can also be cooperative. This could be due to the collusion of the operator undertaking the enrollment, or of other administrative staff. Catching this sort of fraud can be particularly difficult, and relies on strong audit trails and the correct corporate culture where such activities are regarded extremely seriously.

When people are assessing attack likelihoods it needs to be understood that cooperative attacks are likely to have a much higher success rate than covert attacks. For instance, the success rates will obviously be much higher from a fake fingerprint created from a cooperative party, than compared to a covert acquisition (e.g. from a partial latent print). *Although the threat being assessed might be the same, the risk will vary depending on whether the attack is cooperative or covert.*

12.5 Assessing Vulnerabilities and Attack Methods

The evaluation of biometric threats should provide a reliable estimate of the vulnerability for a particular threat using specific technology. It is necessary to ensure that the assessment meets resource constraints, and can flexibly adjust to different biometric modalities in an evolving threat landscape. Ideally, it will also fit in with other security assessment processes.

Biometrics Institute Vulnerability Assessment Methodology

The Biometrics Institute vulnerability assessment methodology provides a principled methodology for assessing the vulnerability of biometric systems to deliberate attacks. A key part of this methodology is to separate the total risk factor into two separate components: *Exploitation Potential*, which relates to properties of the biometric system itself; and *Attack Potential*, which is primarily a function of the capabilities of an attacker [12].

The set of potential threats and threat variants is complex, and a wide variety of factors need to be considered during the evaluation. Each biometric product will have different levels of vulnerability to a range of threats, and each threat is dependent on the attributes of an attacker. Potential threats against a biometric system range from the presentation of artifacts, such as simple printed picture of a biometric, through to the reconstruction of a biometric from stolen biometric templates. The protection profiles established provide one such baseline list of threats [1]. However, the threat list is not static and may continually be expanding as new techniques and materials become available upon which to base attacks.

A general threat list for a biometric system will include many threats that would apply to any complex computer security environment. Given the wide scope of such investigations it is seldom possible to investigate all known threats. One method of dealing with this is to rank the threats in order of how likely the exploit will be for a given system, and investigate those with the highest priority first. The experience of the evaluator to make informed judgments about how the different threats compare, dependending on the nature of the application and what is being protected, is relied upon to ensure this ranking does not yield misleading results.

During testing, vulnerability to a threat is indicated when both the liveness detection is defeated (where it is available) and also high match scores are observed during an attack. A score above the minimum threshold that could be practically set represents a potentially successful attack. The testing process is generally concerned with looking for artifacts that can defeat the liveness tests and obtain the highest similarity scores.

The components of a vulnerability assessment process are to select threats (see Fig. 12.3) and then apply an assessment methodology. This methodology will undertake a testing process for each threat and provide standardized reporting [12].

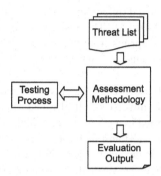

Fig. 12.3 The process of the assessment and reporting of threats.

12.5.1 Attacker Strength

The strength of an attacker is an important consideration when assessing the risk of an attack being successful. Attackers span from a casual impostor or an uneducated criminal with access to information available on the Internet, through to a skilled programmer with access to state sponsorship and full administrator rights. For each increase in attacker strength the likelihood of a system penetration may increase dramatically. However, it can be known a priori that some attacks are impossible without higher level access to the system in order to be able to change settings or inject new information. Similarly, some attacks can be accomplished with very few resources, whilst others require significant skill, time and manufacturing expertize. This difference can be characterized by the level of information that an attacker has, and their level of access to the system. The different levels of information relate to the knowledge of an attacker about mechanisms to attack the system, and span from no particular knowledge through to a detailed knowledge of matching algorithm internals. This could also be put in terms of attacker resources, from bedroom hacker to state sponsored terrorists. The different levels of access an attacker may have span from simple user access (verification, enrollment or both) through to administrator-level access, including source code and re-compilation access [12].

The degree of effort needed to discover, characterize and implement the vulnerability is also a relevant parameter in characterizing the threat. For many simple biometric systems with no functional liveness detection, this discovery effort may be minimal, requiring only hours or days. More complex attacks involving surgery or complex prosthetic construction may take months or years.

12.5.2 The Test Object Approach

One of the most well known and influential biometric vulnerability studies was undertaken by Matsumoto in 2001 [20]. This research involved demonstrating the simple creation of artificial fingerprints from latent fingerprints.[2] Subsequent work by Matsumoto [19] has concentrated on what he calls the *test object* approach to testing.

The test object approach categorizes the classes of artifact attacks using set theory. Consider that from the universe of all physical objects, there is a set that are able to be enrolled and verified in a given system. Within the set of enrolled objects some (in the perfect system - all) will be humans. However, in practice it is highly likely that the universe of all physical objects will also contain some artificial objects that can also be enrolled and verified. There are hence four methods to create artifact attacks (Fig. 12.1)

Enrollment	Verification	Example
human	human	A person whose biometrics are naturally similar enough to be able to pass as someone else.
artificial	human	The use of an artifact to enroll as an impostor, then allowing that impostor later access as themselves.
human	artificial	Co-operative attack where a user allows someone to make an replica of the enrolled biometric for attack purposes.
artificial	artificial	An artifact is used during the enrollment, which can be transferred to another individual later.

Table 12.1 Methods to create artifact attacks [19].

12.6 Vulnerability Mitigations

Mitigations are the steps that can be taken to prevent a specific threat from being exploited. These might involve new sensing mechanisms, changes to the matching algorithm, cryptography, alterations to the environment or usage policy. Developing

---

[2] Latent prints are fingerprint impressions left on a surface after it has been touched.

a mitigation to treat identified vulnerability risks is a vital task for critical systems. Mitigation strategies can be broken into the following categories:

- **Multi-factor Mitigations**: The use of multiple different authentication factors as part of the authentication significantly mitigates against a number of vulnerabilities, as an attacker needs to compromise more than one security mechanism. The use of either a smart-card and/or password/pin combination with a biometric is recommended practice for most secure authentication scenarios. Multimodal biometric solutions may also be used to mitigate risk.[3]
- **Sensor Mitigations**: Different types of sensors can be used to detect artifact attacks and ensure liveness. Examples include the detection of a pulse in fingerprint capture and the detection of reflex action of an iris to light.
- **Signal Processing Mitigations**: Without using any different sensor technologies additional signal processing can be applied to the detection of liveness. Examples include detecting the elasticity of skin as a fingerprint is pressed onto a sensor or noting the deformations expected by a real face compared to a photograph. Signal processing may also be used to check for a replay attack, by examining if a biometric sample is too similar to a biometric seen previously.
- **Behavioral Mitigations**: For biometrics that incorporate a behavioral element such as speech or typing dynamics, mitigation can be applied by asking the user to undertake some behavioral task that can be monitored. Examples include asking a user to speak a random digit string or getting a user to type a random word.
- **Coercion Mitigations**: Where a biometric characteristic has multiple instances such as a fingerprint or iris, one particular instance can be nominated to be used in a "panic" situation. For instance, when the "panic finger" is used it may still allow access but silently raise an alarm. Some biometrics can detect stress through changes to biological signals, such as pitch in voice or increasing pulse rate, however the false alarm rate is often unacceptably high due to the natural variation in such signals.
- **Environmental Mitigations**: The environment in which biometrics are captured can affect its biometric vulnerability. Where biometrics are captured in heavily monitored and policed areas, such as airports, they are more secure than when captured in a private and unmonitored area. Providing surveillance in areas where biometrics are used can greatly enhance the chance of detection of fraud and act as a deterrent to would be attackers.
- **Cryptography Mitigations**: Ensuring the secure transmission of data from each biometric component is vital where the transmission is sent over untrusted networks or insecure communication links. Mitigations may include using a public key infrastructure (PKI) to ensure match decisions are not altered after they are made.

---

[3] It is important to ensure that the multi-factor mitigations are, as much as possible, independent. Two different feature sets from the same physical region (e.g. fingerprint and skin pore, or face and iris) will make covert acquisition of both easier, and hence provide little additional additional security.

- **Tamper Mitigations**: The integrity of the sensor can be both electronically tested and physically secured to ensure that no modifications or substitution have been undertaken. Tamper-proofing might include physically sealing all the internal hardware in resin and using electronic sensors to detect if seals have been broken.
- **Policy Mitigations**: Policy mitigations are those instructions in the use of the system for both operators and users that ensure integrity. One of the most important is ensuring trust in the enrollment process by requiring human supervision, and having policies in place around system administration.
- **Monitoring Mitigations**: By installing an active monitoring system that looks for deviations in normal operational usage, potential fraud relating to system lambs can be determined, or attempted attacks can be determined through an analysis of time series data. Furthermore, the examination of audit logs can potentially reveal patterns of internal fraud.

Appropriate mitigations depend on the system requirements and the value of assets being protected. The trade-off may need to be carefully weighed against practicality since, in some cases, the imposition of a mitigation may have negative side-effects on usability, or may lead to falsely rejecting live input.

12.7 Conclusion

Quantifying the vulnerability of biometric systems and determining appropriate countermeasures is a vital area of research. This chapter has provided an overview of the potential attack points and fraud mechanisms in biometric systems, and an introduction to how they might be assessed for these vulnerabilities.

It is strongly argued that it is important to distinguish between the goal of the biometric matching algorithm, which is to robustly distinguish one person from all others, and the goal of anti-spoofing or liveness techniques, which is to ensure non-human objects are not matched. Both are necessary, but distinct, components of any secure system.

There is a bright future for biometrics systems to enhance and simplify all our interaction with technology. However, in this future the technology will need continued focus on both accuracy and vulnerability.

References

[1] Biometric device protection profile BDPP. http://www.cesg.gov.uk/site/iacs/itsec/media/protection-profiles/bdpp082.pdf (2001)
[2] Communications security establishment certification body canadian common criteria evaluation and certification scheme. http://www.cse-cst.gc.ca/documents/services/ccs/ccs_biometrics121.pdf (2001)

[3] U.S. government biometric verification mode protection profile for basic robustness environments. `http://www.niap.bahialab.com/cc-scheme/pp/pp_bvm_mr_v1.0.pdf` (2001)

[4] Common criteria common methodology for information technology security evaluation: Biometric evaluation methodology supplement BEM. `http://www.cesg.gov.uk/site/ast/biometrics/media/BEM_10.pdf` (2002)

[5] Transcript: Defense department briefing. `http://www.america.gov/st/washfile-english/2002/October/20021017192919ross@pd.state.gov0.9141504.html` (2002)

[6] Episode 59 - crimes and myth-demeanors 2. `http://en.wikipedia.org/wiki/MythBusters_(season_4)#Episode_59_.E2.80.94_.22Crimes_and_Myth-Demeanors_2.22` (2006)

[7] Adler, A.: Sample images can be independently restored from face recognition templates. Electrical and Computer Engineering, 2003. IEEE CCECE 2003. Canadian Conference on **2** (2003)

[8] Boyce, C., Ross, A., Monaco, M., Hornak, L., Li, X.: Multispectral iris analysis: A preliminary study. Proc. Conf. Computer Vision and Pattern Recognition Workshop pp. 51–59 (2006)

[9] Czajka, A., Strzelczyk, P., Pacut, A.: Making iris recognition more reliable and spoof resistant. SPIE The International Society for Optical Engineering (2007)

[10] Daugman, J.: Iris Recognition and Anti-Spoofing Countermeasures. 7th International Biometrics Conference (2004)

[11] Drahansky, M., Lodrova, D.: Liveness detection for biometric systems based on papillary lines. International Conference on Information Security and Assurance, 2008. ISA 2008. pp. 439–444 (2008)

[12] Dunstone, T., Poulton, G., Roux, C.: Update, Biometrics Institute vulnerability assessment project. In: The Biometrics Institute, Sydney Conference (2008)

[13] Faundez-Zanuy, M.: On the vulnerability of biometric security systems. Aerospace and Electronic Systems Magazine, IEEE **19**(6), 3–8 (2004)

[14] Godesberger, A.: Common criteria protection profile biometric verification mechanisms, german federal office for information security (bsi). `http://www.bsi.bund.de/zertifiz/zert/reporte/PP0016b.pdf` (2005)

[15] Harrison, A.: Hackers claim new fingerprint biometric attack. `http://www.securityfocus.com/news/6717` (2003)

[16] Hill, C.: Risk of masquerade arising from the storage of biometrics. Bachelor of science thesis, Dept. of CS, Australian National University (2002)

[17] Kryszczuk, K., Drygajlo, A.: Addressing the vulnerabilities of likelihood-ratio-based face verification. Proceedings of 6th International Conference on Audio-and Video-Based Biometric Person Authentication (AVBPA), T. Kanade and NR (AK) Jain, Eds., vol. LNCS **3546**, 426–435 (2005)

[18] Maltoni, D., Maio, D., Jain, A., Prabhakar, S.: Handbook of Fingerprint Recognition. Springer (2003)

[19] Matsumoto, T.: The test object approach in measuring security of fingerprint and vein pattern authentication systems. In: The Biometrics Institute, Sydney Conference (2008)

[20] Matsumoto, T., Matsumoto, H., Yamada, K., Hoshino, S.: Impact of artificial gummy fingers on fingerprint systems. In: Proc. of the SPIE, Optical Security and Counterfeit Deterrence Techniques IV, vol. 4677 (2002)

[21] Pan, G., Sun, L., Wu, Z., Lao, S.: Eyeblink-based anti-spoofing in face recognition from a generic webcamera. Computer Vision, 2007. ICCV 2007. IEEE 11th International Conference on pp. 1–8 (2007)

[22] Parthasaradhi, S., Derakhshani, R., Hornak, L.A., Schuckers, S.: Time-series detection of perspiration as a liveness test in fingerprint devices. Systems, Man and Cybernetics, Part C, IEEE Transactions on **35**(3), 335–343 (2005)

[23] van der Putte, T., Keuning, J., Origin, A.: Biometrical fingerprint recognition: Don't get your fingers burned. Smart Card Research and Advanced Applications: Ifip Tc8/Wg8. 8 Fourth Working Conference on Smart Card Research and Advanced Applications, September 20-22, 2000, Bristol, United Kingdom (2000)

[24] Schuckers, S.: Spoofing and anti-spoofing measures. Information Security Technical Report **7**(4), 56–62 (2002)

[25] Statham, P.: UK government biometrics security assessment programme, cesg biometrics. http://www.biometrics.org/bc2004/CD/PDF_PROCEEDINGS/bc247a_Statham.ppt (2003)

[26] Thallheim, L., Krissler, J., Ziegler, P.: Body check: biometrics defeated. http://www.extremetech.com/print_article/0,3998,a= 27687,00.asp (2002)

[27] Uludag, U., Jain, A.: Attacks on biometric systems: a case study in fingerprints. Proceedings of SPIE **5306**, 622–633 (2004)

Index